The Growth
Infancy to Adolescence

Third Edition

For Medical, Nursing, Home Science and Anthropology Students

The book in its initial editions and the present 3rd edition
remained the initiative of the
"Health Care and Research Association for Adolescents (adolcare)".

The need to revise what and how to revise was conceptualized by
Prof Anju Seth. I remain indebted to her

Technical help from Ms Priya Bansal MBA (Media Management)
is acknowledged

The Growth
Infancy to Adolescence

Third Edition

For Medical, Nursing, Home Science and Anthropology Students

Editor
KN Agarwal
DCH MD MD(Sweden) FIAP FAMS FNA
President
Health Care and Research Association for Adolescents and
Scientists, Indian National Science Academy

CBS

CBS Publishers & Distributors Pvt Ltd

New Delhi • Bengaluru • Chennai • Kochi • Mumbai • Pune
Hyderabad • Kolkata • Nagpur • Patna • Vijayawada

ISBN: 978-81-239-2530-1

Third Edition: 2015
First Edition: 2003
Second Edition: 2007

Published by Satish Kumar Jain and produced by Varun Jain for
CBS Publishers & Distributors Pvt Ltd
4819/XI Prahlad Street, 24 Ansari Road, Daryaganj, New Delhi 110 002, India.
Ph: 23289259, 23266861, 23266867   Fax: 011-23243014   Website: www.cbspd.com
                                                              e-mail: delhi@cbspd.com; cbspubs@airtelmail.in.
Corporate Office: 204 FIE, Industrial Area, Patparganj, Delhi 110 092
Ph: 4934 4934   Fax: 4934 4935   e-mail: publishing@cbspd.com; publicity@cbspd.com

Branches

- **Bengaluru:** Seema House 2975, 17th Cross, K.R. Road,
  Banasankari 2nd Stage, Bengaluru 560 070, Karnataka
  Ph: +91-80-26771678/79   Fax: +91-80-26771680   e-mail: bangalore@cbspd.com
- **Chennai:** 7, Subbaraya Street, Shenoy Nagar, Chennai 600 030, Tamil Nadu
  Ph: +91-44-26260666, 26208620   Fax: +91-44-42032115   e-mail: chennai@cbspd.com
- **Kochi:** 36/14 Kalluvilakam, Lissie Hospital Road, Kochi 682 018, Kerala
  Ph: +91-484-4059061-65   Fax: +91-484-4059065   e-mail: kochi@cbspd.com
- **Mumbai:** 83-C, Dr E Moses Road, Worli, Mumbai-400018, Maharashtra
  Ph: +91-22-24902340/41   Fax: +91-22-24902342   e-mail: mumbai@cbspd.com
- **Pune:** Bhuruk Prestige, Sr. No. 52/12/2+1+3/2 Narhe, Haveli
  (Near Katraj-Dehu Road Bypass), Pune 411 041, Maharashtra
  Ph: +91-20-64704058-59, 32392277 Fax: +91-20-24300160   e-mail: pune@cbspd.com

Representatives

- **Hyderabad** 0-9885175004   • **Kolkata** 0-9831437309, 0-9051152362
- **Nagpur**   0-9021734563   • **Patna**   0-9334159340   • **Vijayawada** 0-9000660880

Printed at: Magic International Pvt. Ltd., Greater Noida

to

the great pediatrician for our time
Late Shanti Ghosh

List of Contributors

KN Agarwal DCH MD MD (Sweden) FIAP FAMS FNA
President
Health Care and Research Association for Adolescents and Scientist, Indian National Science Academy

DK Agarwal DCH MD
Ex-Professor and Head
Department of Pediatrics
Institute of Medical Sciences
Varanasi

Vijayalakshmi Bhatia MD DM
Professor
Department of Endocrinology
Sanjay Gandhi PGI of Medical Sciences
Lucknow

Anju Seth MD
Professor
Department of Pediatrics
Lady Hardinge Medical College
New Delhi

M Vijayakumar
Additional Professor
Department of Pediatrics
Government Medical College, Kozhikode

PSN Menon MD
Consultant and Head
Department of Pediatrics
Jaber Al-Ahmed Armed Forces Hospital
Kuwait

Sonika Agarwal MD DNB
Pediatric Neurology
Baylor College of Medicine
Houston Tx, USA

Mary Cole MB ChB, MRC Psych
Consultant
Child and Adolescent Psychiatrist
Oxford Health NHS Trust
Salisbury CAMHS; Salisbury
Wiltshire
SP2 8DE, UK

Raj Kumar DCH MRCP(II)
Pediatrician
Metro General Hospital
Noida

AK Bansal
Sr Statistician
University College of Madical Sciences
Delhi

Foreword

Researchers have long accepted that growth patterns in children reflect the overall health of the population to which they belong. This fact means that accurate characterization of child growth is of critical importance, both for assessing individual children, and for assessing parameters related to public health of the society surrounding the children. As a matter of course, environmental and other factors that affect public health change with time, requiring periodic updates of growth data. In that regard, the third edition of *The Growth* presents timely new information.

Chapter 4 has updated information about obesity trends in Indian children and has additional, timely information about risk factors for obesity, such as increased computer time and concomitant decreases in physical activity. The obesity problem is a growing health concern around the world. Its causes have many roots: in affluent groups, sedentary lifestyles and consumption of too many calories contribute to the problem, while in groups of lower socioeconomic status in some nations, poor access to nutritious foods and overconsumption of processed food products and sugary beverages is a significant factor contributing to obesity and poor health. These problems co-exist with malnutrition and the health problems it causes in other low income groups. Such individuals simply do not have access to enough food.

The obesity and calorie/nutrient-poor malnutrition problems are complex, and correct assessment of all of them, both in individuals and at the population level, requires accurate information on the growth of children living in optimal environments. The updates here add important information that will help meet that critical need. They will also be of great benefit to others researching trends in the population of India as a whole or in different groups in India.

An important new emphasis in this edition is the addition of knowledge about brain development in infancy through adolescence. New advances in imaging technologies are helping us map critical processes that occur during human growth and development. In addition, these new technologies have allowed us to view brain processes in real-time and correlate them with normal and pathological states.

Another important new chapter relates to intrauterine growth retardation (IUGR), its causes, and it consequences. IUGR is a serious public health concern, and the chapter has, among other things, solid information on conditions that low birth weight babies are at risk for. References specific to children in India are included.

Overall, the third edition of *The Growth* does an admirable job of gathering detailed information on large number of children from different regions of India and presenting it as a cohesive whole. Clinicians and researchers interested in the health of the Indian population will find much of interest in this volume.

Dr Valerie Natale PhD
The Forgotten Diseases Research Foundation,
Santa Clara, California, USA
www.forgottendiseases.org

Preface to the Third Edition

The third edition has the opportunity to analytically discuss the recent Indian growth studies with the **?** should the earlier study data by us (Agarwal et al 1992; 1994 and 2001) be replaced in favour of the recent ones. Answer is no, as the recent studies deviate in data collection and show **obesity** without significant secular trend in height (see Growth studies in India: Analytical review, Chapter 1). The second **?** should we use the WHO/CDC growth curves, answer is again no. We are ethnically different as shown in comparative growth studies from 55 countries by Natale and Rajagopalan 2014 (BMJ Open 4: e003735). In view of the above important issues the growth curves are redesigned birth to 5 years and 5 to 17 years for girls and 5 to 18 years for boys with the help of Prof Piyush Gupta and Dr AK Bansal, UCMS & GTBH, Delhi (Chapter 1). The sexual development tables for age and sex are also added in appendix.

Recent researches in brain growth and issues to teach our children in advance for protecting them from drugs/smoking/alcohol/sex issues (Time Magazine 2004; 10th May—Secrets of the Teen Brain and July 2013). In Chapter 2 brain in adolescence is discussed along with nutrient needs. The current issues on brain growth and plasticity of brain are discussed in two separate chapters (by a Pediatric neurologist from the USA and Child–Adolescent Psychiatrist from the UK).

The contributing colleagues from Lady Hardinge Medical College, New Delhi; SGPGI Medical Sciences, Lucknow; former AIIMS professor currently in Beirut have churned the recent issues in their field.

I am confident medical, anthropology, nutrition, home science and school teachers will benefit from the book.

KN Agarwal
Editor

Preface to the First Edition

Human growth is an important subject of study for health care providers and pediatricians to assess, monitor and treat if there is a treatable cause of growth failure. Home science students in the field of 'Child development' and 'Nutrition' need knowledge on growth. Anthropologists use human growth for establishing ethnic variations and secular trend. Studies from the USA and England have concluded that the Indians born and nurtured in these countries have different growth pattern as compared to Caucasians. Attempts to establish internationally usable growth indices like 'Body mass index' has not provided smooth curves during adolescence (9–18 years of age).

There is no handbook available to describe "Human growth" from infancy to adolescence for the students of medicine, child development, nutrition and anthropology. This book on growth is based on our clinical and research studies on growth, infancy to adolescence, in children living in affluent and in rural endemic undernutrition. The book gives definition of various growth indices, cutoff for identifying variations in growth for the Indian children. There is emphasis to use data from Indian affluent children as standard for growth assessment and diagnosis of severity of malnutrition. For the first time percentile tables and growth curves on Indian affluent children in relation to age are provided. For adolescent boys and girls growth curves in relation to sexual maturity rating (SMR) are presented for faltering in growth. These growth curves and percentiles for ages and sexual maturity will not only provide standards for measuring growth but also help to diagnose growth variations and identify those 'at risk of undernutrition' and 'obesity'. These data sets will also serve to study secular growth trend in coming decades.

The chapter on Puberty and Adolescence will help in understanding the growth needs of this age group. The adolescent children who have been neglected are presently under special attention of the international community. The chapter on Skeletal Growth offers simple method for assessment of age. We hope this book shall be of use to Pediatricians, Nutritionists, Anthropologists, General Practitioners and Paramedical Groups including Nursing staffs.

We wish to thank the subject of these database, i.e. children and their parents for helping in collection of such huge data. We also like to thank principals and teachers of these schools for their cooperation in undertaking the study during busy school hours. We are thankful to Shri KB Kapoor of NBT for encouraging us to write this book.

We will appreciate criticism and suggestions to improve the book.

KN Agarwal
DK Agarwal

Contents

The Growth and Its Assessment

KN Agarwal

DEFINITION

The term **'growth'** refers to increase in body size or mass. Every organ of the body participates in growth. Human growth stretches from conception (intrauterine) to first two decades of life. The fastest pace is between conception and birth. The growth curve exhibits (i) rapid growth between 20th week, intrauterine to early infancy; (ii) a long phase of slow growth in childhood, and (iii) secondary acceleration during adolescence (Fig. 1.1).

- Human growth from infancy to maturity involves great changes in body size and appearance, including the development of the sexual characteristics. The growth process is not a steady one: as sometimes growth occurs rapidly, at others slowly. Individual patterns of growth vary widely because of differences in heredity and environment. Children tend to have physiques similar to those of their parents or of earlier forebears; however, environment may modify this tendency. Living conditions, including nutrition and hygiene, have considerable influence on growth.

- Thus 'Growth' is a continuous process commencing at conception and progressing at a varying pace till its completion about

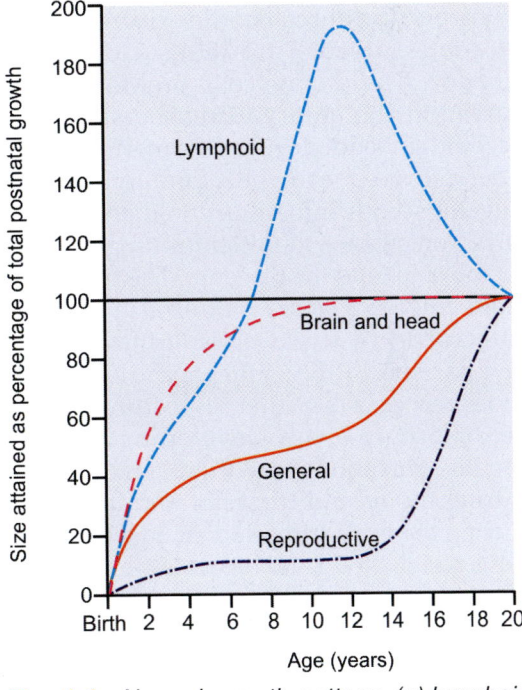

Fig. 1.1: *Normal growth pattern: (a) lymphoid tissue; (b) brain; (c) general, and (d) reproductive*

2 decades later, with closure of epiphysis. The process of 'Growth' is accompanied with increase in body size and or mass at varying rates. It is multi-factorial and complex, still remarkably predictable. Boys and girls grow differently and each child has his or her distinct growth pattern.

INTRAUTERINE GROWTH

Fertilization occurs when the sperm successfully enters the ovum's membrane. The genetic material of the sperm and egg then combine to form a single cell (zygote) and the germinal stage of prenatal development commence (fertilization-early embryo-implantation). The germinal stage lasts for about 10 days of gestation.

The zygote contains a full genetic material and develops into the embryo. Prior to implantation, the embryo remains in a protein shell, the zona pellucida, and undergoes a series of cell divisions. A week after fertilization the embryo still has not grown in size, but hatches from its protein shell and adheres to the lining of the mother's uterus, inducing a decidual reaction. The uterine cells proliferate and surround the embryo thus causing it to become embedded within the uterine tissue. The embryo, meanwhile, proliferates and develops both into embryonic and extra-embryonic tissue, the latter forming the fetal membranes and the placenta. The transition from embryo to fetus is arbitrarily defined as occurring 8 weeks after fertilization. In comparison to the embryo, the fetus has more recognizable external features, and a set of progressively developing internal organs. Maximum growth in length occurs during the second trimester and in weight during the third trimester. At 20 weeks, the fetus weighs 460 g and has a length of 19 cm. These increase to 900 g and 25 cm at 25 weeks. Gestational age in relation to length and weight are given in Table 1.1 (Doublet et al 1997; Hadloek et al 1992).[1, 2]

The data from Indian studies and the USA for birth weight in relation to gestation are given in Table 1.2.

- *Fetal growth* is critical to a person's eventual height. Before birth, the key measures assessed by ultrasound are crown-rump and crown-heel length.

Table 1.1: *Gestational age in relation to length and weight*

GA wk	Length cm	Weight g
30 weeks	39.9 cm	1319 g
31 weeks	41.1 cm	1502 g
32 weeks	42.4 cm	1702 g
33 weeks	43.7 cm	1918 g
34 weeks	45 cm	2146 g
35 weeks	46.2 cm	2383 g
36 weeks	47.4 cm	2622 g
37 weeks	48.6 cm	2859 g
38 weeks	49.8 cm	3083 g
39 weeks	50.7 cm	3288 g
40 weeks	51.2 cm	3462 g
41 weeks	51.7 cm	3597 g
42 weeks	51.5 cm	3685 g
43 weeks	51.3 cm	3717 g

GA: gestational age

Later on serial measurements of biparietal and occipito-frontal diameters of skull are used to assess fetal growth.

- The fastest growth rate for a human is during embryonic life (rate being 50–60 cm/yr).

- The growth of the embryo and fetus is mainly mediated by maternal nutrition and by growth factors such as fibroblast and epidermal growth factors, transforming growth factors alpha and beta, insulin, and insulin-like growth factors (IGF-I and IGF-II).

- The growth hormone only begins to play a role in growth in the final weeks before birth.

Physical Stages

- Stages in prenatal development— **Embryo**-fertilization—8 weeks after fertilization

- **Fetus**—10th week of pregnancy—birth neonate (newborn) (0–30 days)

Table 1.2: *Birth weight data compiled in relation to gestational age*

Gestation (wk)	Weight (g)					
	Ghosh et al.[3] 1971	Mittal et al.[4] 1976	Bhatia et al.[5] 1981	Agarwal et al.[6, 7] 2000, 2002	Lubchenco et al.[8] 1963, 1966	
	Delhi	Ludhiana	Varanasi		Denver, USA	
			Urban	Rural	Boy	Girl
30	1326 ± 279	1476 ± 322	1155	–	–	–
32	1608 ± 314	1767 ± 436	1575	–	1760	1675
33	1941 ± 562	2039 ± 400	1755	–	–	–
34	2052 ± 616	2156 ± 424	1955	–	–	–
35	2205 ± 629	2231 ± 424	2145	2275	–	–
36	2421 ± 558	2479 ± 435	2345	2550	2745	2630
37	2691 ± 464	2601 ± 421	2525	2600	–	–
38	2760 ± 442	2791 ± 401	2690	2650	–	–
40	2895 ± 460	2974 ± 425	2865	2670	3290	3160
42	2927 ± 441	3017 ± 362	2830	2670	3310	3210

Commonly used Nomenclature

Preterm: Live birth delivered before 37 weeks from the first day of the last menstrual period.

Extremely low birth weight (ELBW) <1000 g

Very low birth weight (VLBW) <1500 g

Low birth weight (LBW) <2500 g. These include preterm appropriate for gestational age (AGA), or small for gestational age (SGA) and full term SGA babies.

GROWTH IN INFANCY

(BIRTH TO FIRST YEAR OF LIFE)

Newborn period is defined as the first 4 weeks after birth. In full term babies (gestation 37–42 wk) birth length is around 48–50 cm, weight ranges between 2.7 and 4.6 kg average being 3.4 kg, and head circumference is around 34–36 cm. The infant loses weight during the first few days after birth, but regains it by 10–15 days of life. The pattern of gain in weight, length and skull circumference during the first year for affluent Indian infants is given in Table 1.3.

Table 1.3: *Growth in infancy [Agarwal et al. (1994)[9]]*

Age (months)	Weight (kg)	Length (cm)	Skull (cm)	Weight* gm/day	Length* cm/month	Skull* cm/month
Birth	3.1–3.4	50	3.4–3.5			
3	5.5	60	40	25–30	3.5	2.0
6	7.7–8.0	65.5–68.0	42–43	20	2	1
9	8–9	70–72	44–45	15	1.5	0.5
12	9–10	73–75	46–47	12	1.3	0.3

*Velocities

CHILDHOOD GROWTH

Common terms used in child growth period
- **Toddler** (1–3 years)
- **Middle childhood** (4–12 years)
- **Preadolescent**—The child in this and the previous phase are called *school child* (*school boy* or *school girl*), when still of primary school age (10–12 years).
- **Adolescence and puberty**
- *Girls* generally enter puberty between ages 9 and 12 years of age.
- *Boys* usually enter puberty at ages 12–15 years of age.
- *Adulthood* (20+ years) the progressive development of a living thing, especially the process by which the body reaches its point of complete physical development.

Weight

In general, the birth **weight** of the average baby doubles in 5–6 months and triples by the end of the first year.

The **general** guidelines that are usually given for growth are:
- A baby loses 5–10% of birth weight in the first week and regains this by 2–3 weeks.
- Birth weight is doubled by 4–6 months and tripled by 12 months.
- Birth length increases 1.5 times in 12 months.
- Birth head circumference increases by about 10 cm in 12 months.
- At the end of the second year of life birth weight quadruples and then there is a steady increase of 2–2.75 kg (4.4–6 lb) each year until the child reaches puberty, at which time there is a period of rapid growth in weight and height (*see* Fig. 1.1).
- Weight-for-age is usually used to monitor growth. It is particularly useful in small infants who normally gain weight fast. Normal weight gain suggests that the infant is healthy and growing normally. *Failure to gain weight normally is often the earliest sign of illness or malnutrition (i.e. undernutrition).*

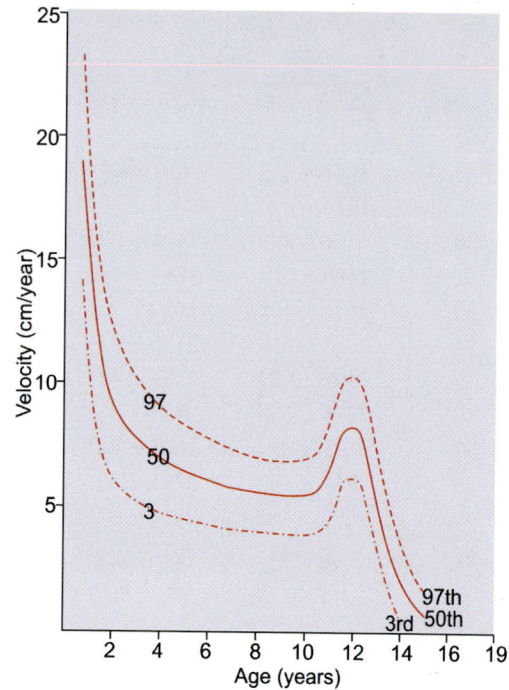

Fig. 1.2: *Height velocity percentiles for age. Note the deceleration with second spurt in adolescence [Tanner and Whitehouse[10]]*

Height

- Is the best index of measuring linear growth (stature) as height reflects growth over a longer period than does weight. *During 1st year of life 25 cm; 2nd year = 12.5 cm; 3rd year = 7.5–10 cm; and 7cm/year at 3–4 years, and 5-year onwards = 5 cm/yr until puberty.*
- Growth in **height** occurs as a result of maturation of the skeleton. When the long bones have reached maturity at about age 18, linear growth stops.

Head Circumference

Head circumference can be used to assess brain growth in children mainly under 2 years, as during this period brain growth is very rapid (*at birth, the brain of the infant is 25% of the adult size. At the age of one year, the brain has grown to 75% of its adult size and to 80% by age three, reaching 90% by age seven*).

The head circumference increases at the rate of 2 cm per month during the first 3 months of life, 1 cm a month in next 3 months and 0.5 cm per month during the later 6 months of infancy. There is 12 cm increase in head circumference from birth to first birthday. Thus, the skull circumference is 34–35 cm at birth in full term baby, grows to 40 cm at 3 months, 46–47 cm on the first birthday and 48 cm at 2 years of age.

Fontanels (Fig. 1.3)

Anterior (frontal) fontanel (AF): A diamond shaped open gap is situated in midline at the junction of the coronal and sagittal sutures (size around 2 × 2 cm) closes by 18 months (range 10–18 months). Small or closed AF is a warning sign, i.e. microcephaly. Large AF may be present in hydrocephalus, hypothyroidism, down syndrome, achondroplasia, osteogenesis imperfect and mucopolysaccharidosis.

Posterior (occipital) fontanel (PF) placed between the intersection of the occipital and parietal bones, closes within 6–8 weeks of life, an open PF in later life is noted in congenital hypothyroidism.

Growth in childhood: As growth slows, children need fewer calories and parents may notice a decrease in appetite. Two-year-old child can have very erratic eating habits that sometimes make parents anxious. It seems as though some children eat virtually nothing yet continue to grow and thrive. Actually, they eat little one day and then make up for it by eating everything insight the next day.

Most children aged six to eight years will

- Experience slower growth of about 5 cm and 2–3.5 kg per year.
- Grow longer legs relative to their total height and begin resembling adults in the proportion of legs to body.
- Develop less fat and grow more muscle than in earlier years.
- Increase in strength.
- Lose their baby teeth and begin to grow adult teeth which may appear too big for their face.
- Use small and large motor skills in sports and other activities.
- **Delayed growth and development**—defined as deviation from age group norms.
- **Risk for disproportionate growth**—defined as being at risk for growth above the 97th percentile or below the 3rd percentile for age, or crossing two percentile channels.

Growth and Development

- **Growth** refers to specific body changes and increases in the child's size (such as: height, weight, head circumference, and body mass index). These size changes can easily be measured (*see* growth curves).
- **Development** typically refers to an increase in complexity (a change from simple to more complex), involves a progression along a continuing pathway on which the child acquires more refined knowledge, behavior, and skills. The sequence is basically the same for all children, however, the rate varies.

Frontal fontanel

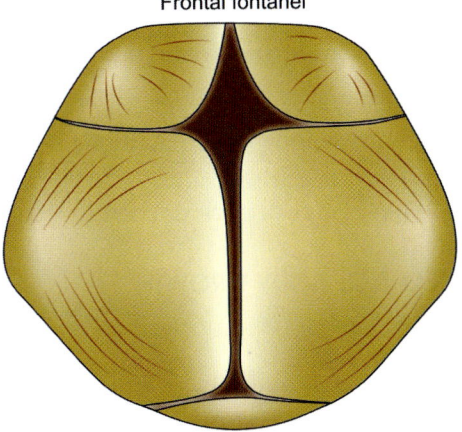

Occipital fontanel

Fig. 1.3: *Frontal and occipital fontanels*

- **Similarities in growth**
- Growth proceeds from the head downward and from the center of the body outward.
- Children gain control of the head and neck first, then the arms and finally the legs.
- At birth, the brain, heart, and spinal cord are fully functioning to support the infant.
- As children grow, the arm and leg muscles develop followed by the finger and toe muscles.
- **Differences in growth**
- Children differ in their growth. Some children are taller, some shorter.
- Some children are smaller, while others are larger.
- These differences are completely normal. Normal growth is supported by good nutrition, adequate sleep and regular exercise.

Key Box

- *Growth* is defined as specific body changes and increases in the child's size. Growth proceeds from the head downward and from the center of the body outward. Children differ in their growth.
- Growth is a fundamental characteristic of childhood
- Despite being influenced by many factors, it remains remarkably predictable
- Normal growth is an indicator of optimum health
- Deviation from the normal pattern is indicative of a pathological process
- Periodic assessment facilitates early detection of growth faltering which may be the first manifestation of under/over nutrition, infection/disease
- *Development* typically refers to an increase in complexity, a change from relatively simple to more complicated. Development usually involves a progression along a continuous sequential pathway on which the child acquires more refined knowledge, behaviors, and skills. The sequence is basically the same for all children; however, the rate varies.

- Children do not grow at perfectly steady rates throughout childhood.
- Children will experience weeks or months of slightly slower growth followed by growth spurts
- Difference in the amount of growth can be a source of self consciousness for some children. It is important to help the children in your care understand that these differences are normal, that each child is special, and to help children develop a sense of self acceptance.

ADOLESCENT GROWTH

The second growth spurt after the one observed during fetal life and early infancy occurs during adolescence with gains of 20% of total stature and 40–50% adult weight. During pubertal period boy gains 29–30 cm in height and 28 kg weight and for girl corresponding figures are 24–25 cm and 25 kg weight respectively.

ADULT STATURE

A child's genetic potential for height can be assessed by the mid-parental height (MPH) which is calculated by the formulae given below:

MPH for girls (cm)

$$= \frac{(\text{Father's height} - 13) + \text{Mother's height}}{2}$$

or $= $ Average of parental height $- 6.5$ cm

MPH for boys (cm)

$$= \frac{(\text{Mother's height} + 13) + \text{Father's height}}{2}$$

or $= $ Average of parental height $+ 6.5$ cm

PATTERNS OF BODY GROWTH

In Fig. 1.1, see four important growth components in relation to age:

a. Lymphoid tissue (immunity);
b. Brain and head (intelligence and cognition);
c. General growth, and
d. Reproductive development

a. *Lymphoid:* The growth of lymphoid tissue is the highest during mid-childhood, e.g. enlarged tonsils and lymph nodes, maximum being at 8–9 years of age, and later decreases in size. This helps in development of immune system.

b. *Brain (head circumference-brain size):* Brain growth occurs very rapidly during 20th weeks of intrauterine life to 20th month of infancy. Although brain cell formation is almost complete before birth, brain maturation continues after birth. The brain of the newborn is not yet fully developed. It contains about 100 billion brain cells that have yet to be connected into functioning networks. But brain development up to age one is more rapid and extensive than was previously realized. At birth, the brain of the infant is 25% of the adult size. At the age of one year, the brain has grown to 75% of its adult size and to 80% by age three, reaching 90% by age seven. The influence of the early environment on brain development is crucial. Infants exposed to good nutrition, toys, and playmates have better brain function at age 12 than those raised in a less stimulating environment. The rapid brain growth is reflected by an increase in head circumference. Recent studies show that it is around 25 years of age that brain development completes.

c. *General body:* The general body growth is rapid during fetal life and first 1–2 years of age. The growth velocity slows later during mid-childhood and accelerates once again during puberty. The limbs and arms grow faster than the trunk, so that body proportions undergo marked variation as an infant grows into an adolescent. There is a rapid deceleration in the rate of growth as the child grows beyond infancy. The lowest velocity point is reached around 10 years in girls and 12 years in boys.

Thereafter, the adolescent growth spurt supervenes attaining rapid growth velocity which finally decelerates as the growth completes (Fig. 1.4).

d. *Reproductive (sexual development)* grows at different rates around 9–11 years in girls and 11–13 years in boys. The sexual development is complete by 19–20 years of age.

FACTORS REGULATING LINEAR GROWTH

The process of gain in stature (height) ceases with closure of epiphysis. However, hair, skin and gastrointestinal cell lining continue to grow. The process of growth is largely determined genetically, but is also influenced by environment. The various factors affecting growth are as follows:

a. *Genetic influences:* Racial differences are an important cause of variations observed in human growth (Table 1.4).

b. *Environmental influences:* Any insult especially one which is prolonged or severe (e.g. poverty), particularly during the periods of rapid growth, leads to irrevocable damage to growth and development process.

c. *Factors influencing the 'Intrauterine growth'* gets influenced by smoking,

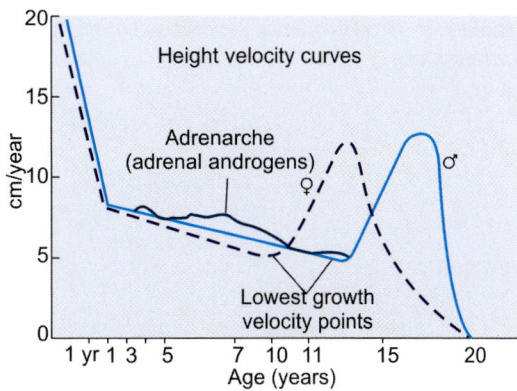

Fig. 1.4: *Rapid deceleration of growth velocity in prepubescence. Girls are advanced by 2 years over boys [modified—Tanner et al. 1965][11]*

alcohol intake and in nutrition deprivation during pregnancy. The examples of nutrition deprivation/poor pregnancy nutrition were observed during World War II. The experience of *world war II* from Leningrad (USSR) famine (period 16 months) do suggest that *pre-pregnancy undernutrition when faced with acute malnutrition in pregnancy:*

- Birth weight fell by 530 g
- Exposure in second trimester resulted in 50% with birth weight <2500 g.

In contrast, Dutch famine was imposed on a *previously well-nourished population*. It reduced birth weight by 327 g and corresponding figure for those with weight <2500 g was 9% only.

These studies convince us that pre-pregnancy nutrition is important for health of child and mother.

Secondly "Rural pregnancy" data from Varanasi (1987–1993)[6,7] showed that:
- 27.4% babies at birth were small in size (<2500 g),
- 6.6% were preterm births (born before full term), and
- Only 8.2% weighed >3000 g (normal birth weight being around 3.3 kg).

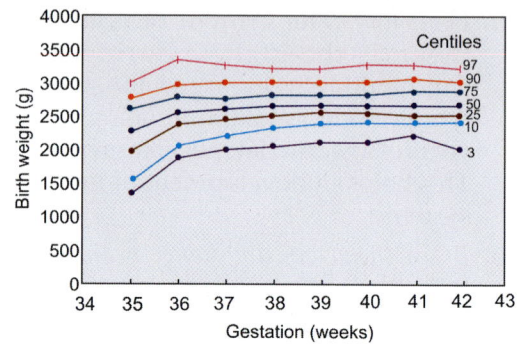

Fig. 1.5: *Birth weight percentiles for gestation (rural Varanasi n = 3700), live births: 7.2% were <2250 g and 27.4% <2500 g. The weekly birth weight increments in gestation 36–42 weeks were 5–53 g, only). Permission Ed. (Prof. P Gupta) Indian Pediatrics 2002; Agarwal S et al.[7]*

More births with weight <2500 g, were due to **maternal undernutrition** resulting in *fetal growth arrest*. These women in later pregnancy (during 36–42 weeks of gestation) showed weekly weight gain of 5–53 g, only (Fig. 1.5). In contrast women in Ludhiana gained >100 g per week.[4] This was due to poor nutrition in pre-pregnancy and pregnancy period.

d. Emotional deprivation—affects physical growth as well as mental functions (development).

e. *Thyroxin and growth hormones (GH)* are the key hormones promoting growth. Absence of either severely retards growth. Much higher GH levels are observed at the neonatal and early childhood periods and then again at puberty as compared to adults. At puberty, sex steroids work synergistically with growth hormone (GH) to bring about the growth spurt. However, higher level of testosterone is needed as compared to *estrogen* to bring about similar degree of growth. Thus, in girls, pubertal growth spurt occurs earlier. Insulin, prolactin and pituitary gonadotropins also play a role in growth promotion.

Table 1.4: *Growth of body proportions (genetically determined)*

a. Size
  - Afro-Americans are taller than Americans
  - Asians are smaller than Europeans, irrespective of similar environment

b. Shape
  - Aborigines – Longest legs
  - Africans – Longer legs and arms
    - Narrower hip as compared to shoulder
  - Asians and – Comparable in body portion
    Europeans
  - Japanese – Large trunk as compared to legs

f. *Chronic diseases:* Many chronic diseases can retard growth. Notable among them are asthma, hemolytic anemias, congenital heart disease, malabsorption states, renal tubular acidosis and renal failure. GH levels are normal, and often elevated in these disorders. However, IGF-I (insulin like growth factor-I) levels tend to be low in these situations. It is worthwhile to point here that GH exerts its growth promoting effects primarily through generation of IGF-I.

g. *Growth variation in response to physiological needs:* Alaskans (Tundra) are shorter, stockier, bulkier in physique with shorter appendages. This is because diversion of nutrients from growth to heat production stunts their growth. They also have wider and larger skulls to mitigate heat loss.

h. *Ethnic growth variations:* Variations in growth are observed among different populations. While average height and weight of pre-adolescent well-to-do children tend to be similar, clear ethnic differences appear during adolescence. Thus, Indian and other Asian children who are at 50th centile of NCHS and the recent WHO[12] Standard till 5 years of age, become close to the 25th centile at the end of adolescence. Thus, the WHO recommends that NCHS standards should not be used to evaluate growth of Asian children beyond 10 years of age. Instead, local standards derived from well-nourished children should be used for growth comparison. For India, standards derived by Agarwal et al[9, 13, 14] are suitable. The NCHS—WHO standards are however needed for comparison of nutritional status among different populations across the world. WHO in April, 2006 (Study period 1997–2003)[12] generated longitudinal growth data from birth to 24 months and a cross sectional data on children aged 18 to 71 months. The data were collected from Brazil, Ghana, India, Norway, Oman and USA by selecting healthy children living under conditions likely to favor the achievement of their full genetic potential. However, in view of ethnic and geographical differences WHO standards[12] are debated recently[15, 19] in favor of local standards.

REGULATION OF GROWTH

Glands and growth. The regulators of growth are the Endocrine glands, they are subject to hereditary influence. The pituitary gland secretes growth hormone, which controls general body growth, particularly the growth of the skeleton, and also influences metabolism. In addition to influencing growth directly, the pituitary gland has a central role in regulating the other endocrine glands. These other glands in turn control many body functions, and they secrete the various hormones that directly regulate metabolism.

a. *Infancy:* Growth in various phases of life is regulated by different mechanisms.

From 20 wk of gestation to six months postnatally the growth is primarily under the control of brain/hypothalamus and is nutrition dependent.

b. *Childhood:* Six months of age until the prepubertal period, GH (growth hormone) and thyroxine are the key hormones controlling growth. The childhood component of growth adds about 70 cm to height.

During late childhood adrenal androgens also influence growth, particularly in boys—**Adrenarche**

c. *Adolescence:* Pubertal growth is regulated by both GH and sex hormones. Thyroxine continues to play an important role in pubertal growth.

Variations in growth rates: The growth of different individuals varies a great deal. It should be remembered that the rate of growth we call "normal" is really only an average rate. There is a wide range of growth rates, almost all of them quite normal. Of the children of a given sex and age, only about two-thirds will have physical measurements that fall close to the average.

Growth monitoring is the regular measurement of a child's size in terms of length, weight and brain size as head circumference in order to document growth with age. The child's size measurements must then be plotted on a growth chart. This is extremely important as it can detect early changes in a child's growth. Both growing too slowly or too fast may indicate a health problem.

Growth charts (curves) are used to measure growth. The distance curve is a measure of size over time; it records height/weight as a function of age and gets higher with age (*see* growth curves for head circumference, height, weight and BMI). Growth charts consist of a series of percentile curves that illustrate the distribution of selected body measurements in children.

Pediatricians, nurses, and parents use these to track the growth of infants, children, and adolescents.

Percentiles (centile) describe the frequency distribution of anthropometric parameters like weight, height, skull circumference, BMI etc. In 50th percentile is the average (median) line for the given population. Describes the percent of children expected to be on or below that line, e.g. 50th centile means that 49% of the observations are below and 50% above that observation. A child's growth parameters may be on the centile line or between two centile lines. Conventionally, for all parameters, 3rd and 97th percentiles are the lowest and highest 94% of observations. (This coincides with values ± 2SD from the mean.) The 50th centile (median value) for any age indicates that for the given parameter deviation below the center is just the same as the deviation above (Fig. 1.6). In case of normal Gaussian distribution mean and 50th centile (median) values are the same.

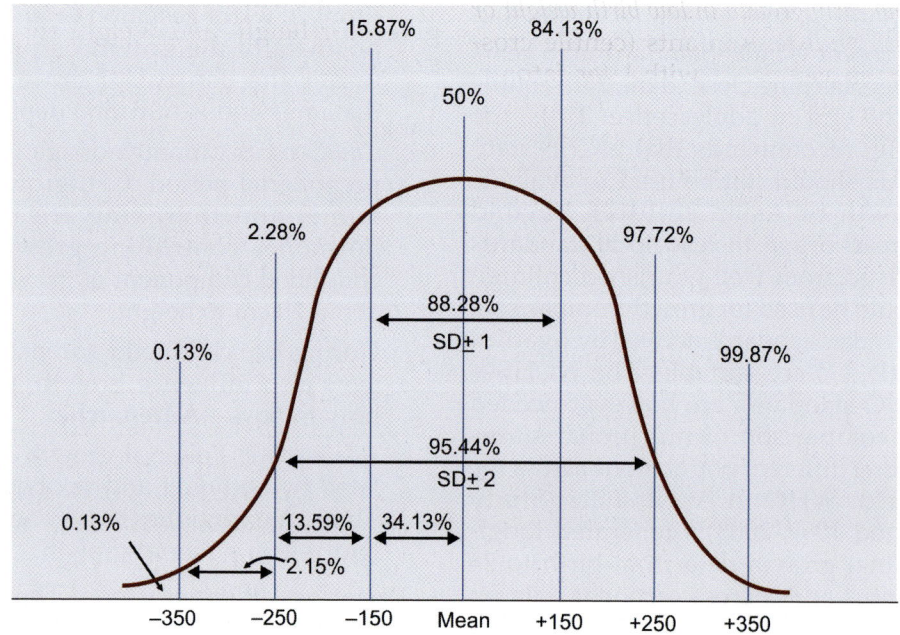

Fig. 1.6: *Percentile distribution in relation to standard deviations (SD)*

- Any child with parameters below or above these limits 3rd and 97th centile or those who cross percentiles needs careful evaluation.

Examples

- If height and weight consistently are on the 60th percentile line until a child is 5 years old, then the height has dropped to the 30th percentile at age 6, that might indicate that there's a growth problem (catch down—retardation of growth) because the child is not following his or her previous growth pattern—this indicates disease.

- Boy with height in 40th percentile and weight in the 85th percentile. (He is taller than 40% of kids his age, but weighs more than 85% of kids his age). There might be a health problem (overweight/obesity). On the other hand, if he is in the 85th percentile for height and weight and follows that pattern consistently over time, that usually means that he is a normal child, just larger than average.

- *Rapid early growth in low birth weight or healthy full term* infants (centile crossing) is associated with later fatness, obesity, blood pressure, cholesterol and insulin resistance—the key risk factors for CVD (Singhal et al).[16] Buyken et al[17] showed that breastfeeding reduces these risks. Drop in 2 centiles suggests illness and demands careful examination.

How to plot in the growth curve? (*All the points on the growth chart should be marked only as dots and not circles around the dot*).

1. The first step is to find the right growth chart. For example, if we are going to find the percentile for a 2-year-old boy, so we will use the growth chart for boy's birth to 36 months. Write child's name, date of birth and date of examination on the top of the curve.

2. Next, find your child's age at the bottom of the chart and draw a vertical line (a straight line up and down) on the growth chart. Now find your child's weight on the left hand side of the chart, and draw a horizontal line (a) straight line from side to side). (*Do not have to really physically draw a line on the growth charts. If you really do that each time, your growth chart will look very messy and will be hard to read. Instead, just imagine where the line should be or draw a light line with a pencil that you can later erase*).

3. Find the spot where these two lines intersect or cross each other. Find the curve that is closest to this spot and follow it up and to the right until you find the number that corresponds to your child's percentile.

Agarwals' Growth Curves (Source Agarwal et al[9, 13]) (Figs 1.7 to 1.10)

The Indian affluent growth data are in appendix Tables 7–16 (height, weight and skull circumference).

WHO growth charts birth to 2 years are given for height and weight for use, as practiced in the USA (Figs 1.11 and 1.12). WHO/NCHS growth data are in appendix Table 1–6.

Adult Stature

To find out the target height for the child—measure the parents and make a note of their heights on the chart. Calculate the child's target height and plot it at 18 years for boys and 15 years for girls and mark it with an arrow on the growth chart. This represents the child's projected height and his present height centile can be judged by tracing a line backward from this target height to child's current height. The target range is produced by plotting two points 8.5 cm above and below the target height and this represents the 3rd and the 97th centile for that child. Taking those two points

Name: _____   DOB: _____   Record No.: _____

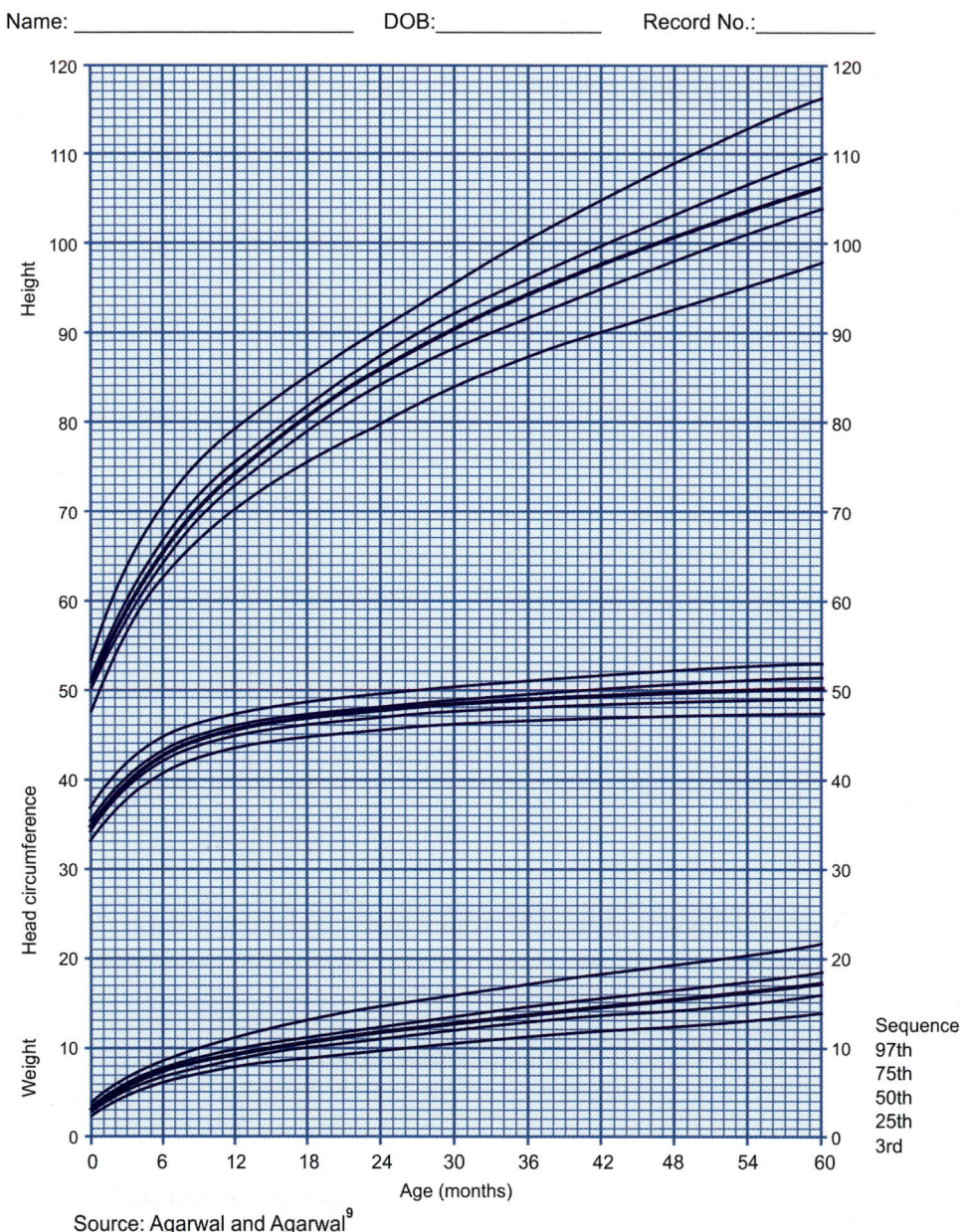

Source: Agarwal and Agarwal[9]

Fig. 1.7: *Height in cm, weight in kg and head circumference in cm percentile curves (Boys 0–5 years)*

Name: _____ DOB: _____ Record No.:_____

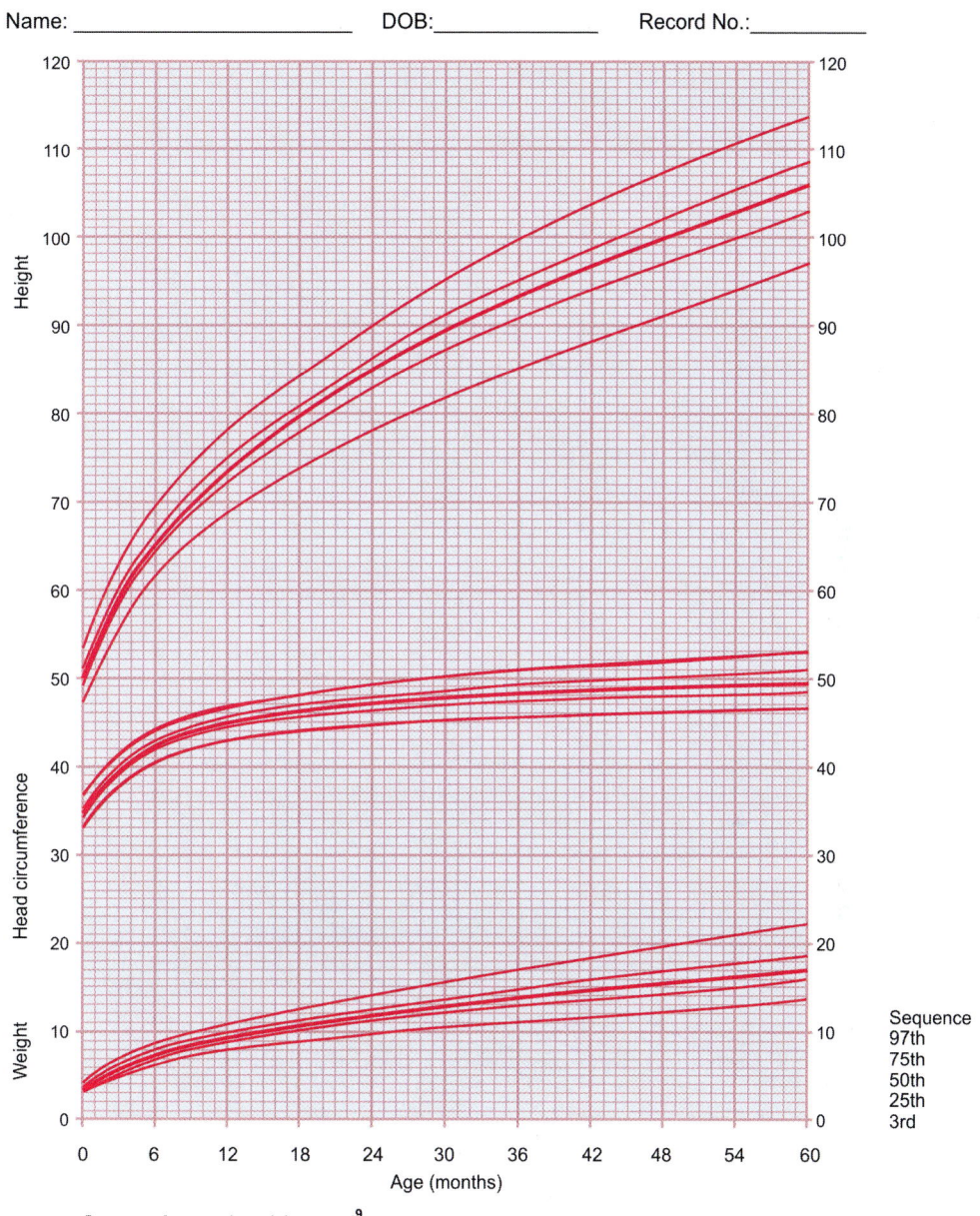

Source: Agarwal and Agarwal[9]

Fig. 1.8: *Height in cm, weight in kg and head circumference in cm percentile curves (Girls 0–5 years)*

Name: _____   DOB:_____   Record No.:_____

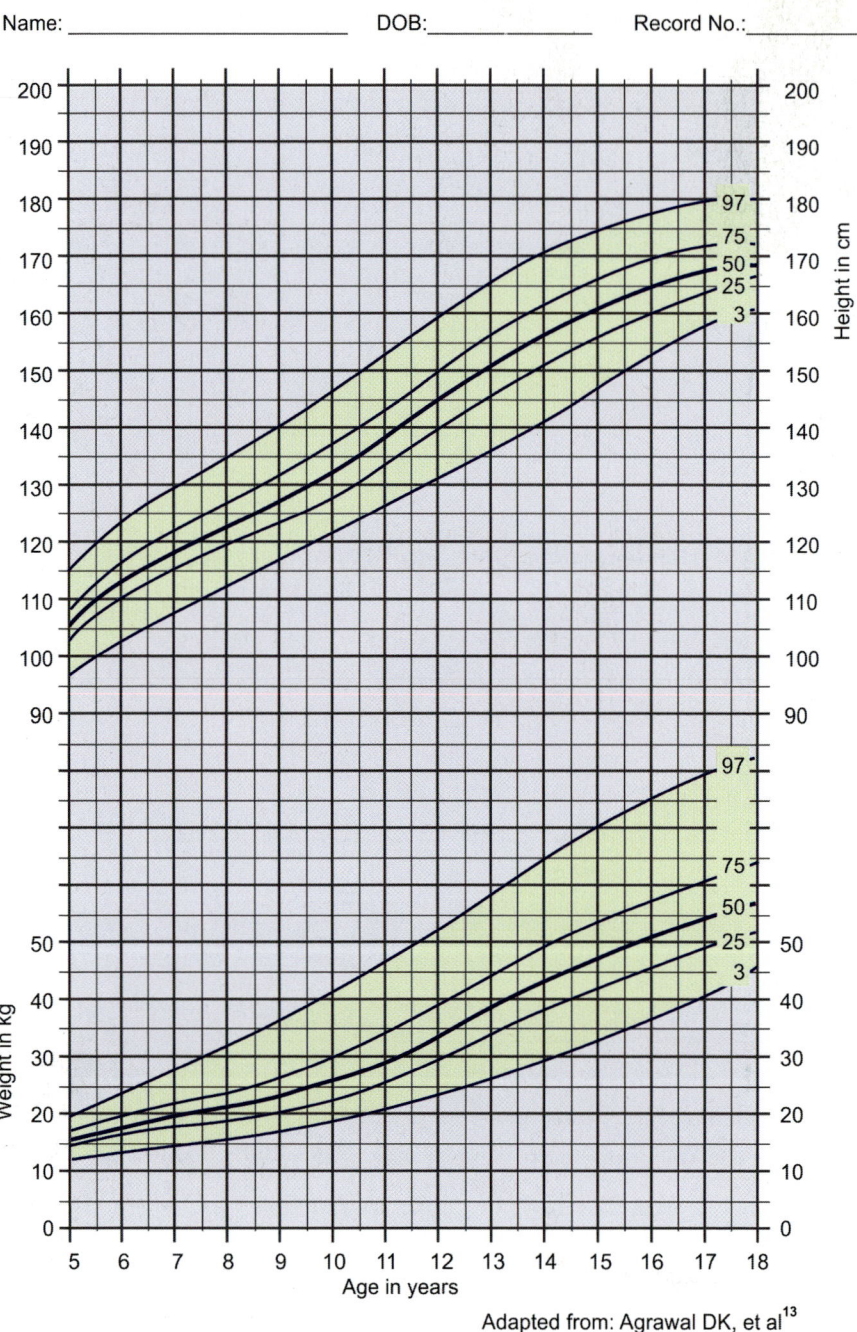

Adapted from: Agrawal DK, et al[13]

Fig. 1.9: *Growth chart for Indian children weight-for-age and height-for-age percentiles (Boys 5–18 years)*

Name: _____ DOB:_____ Record No.:_____

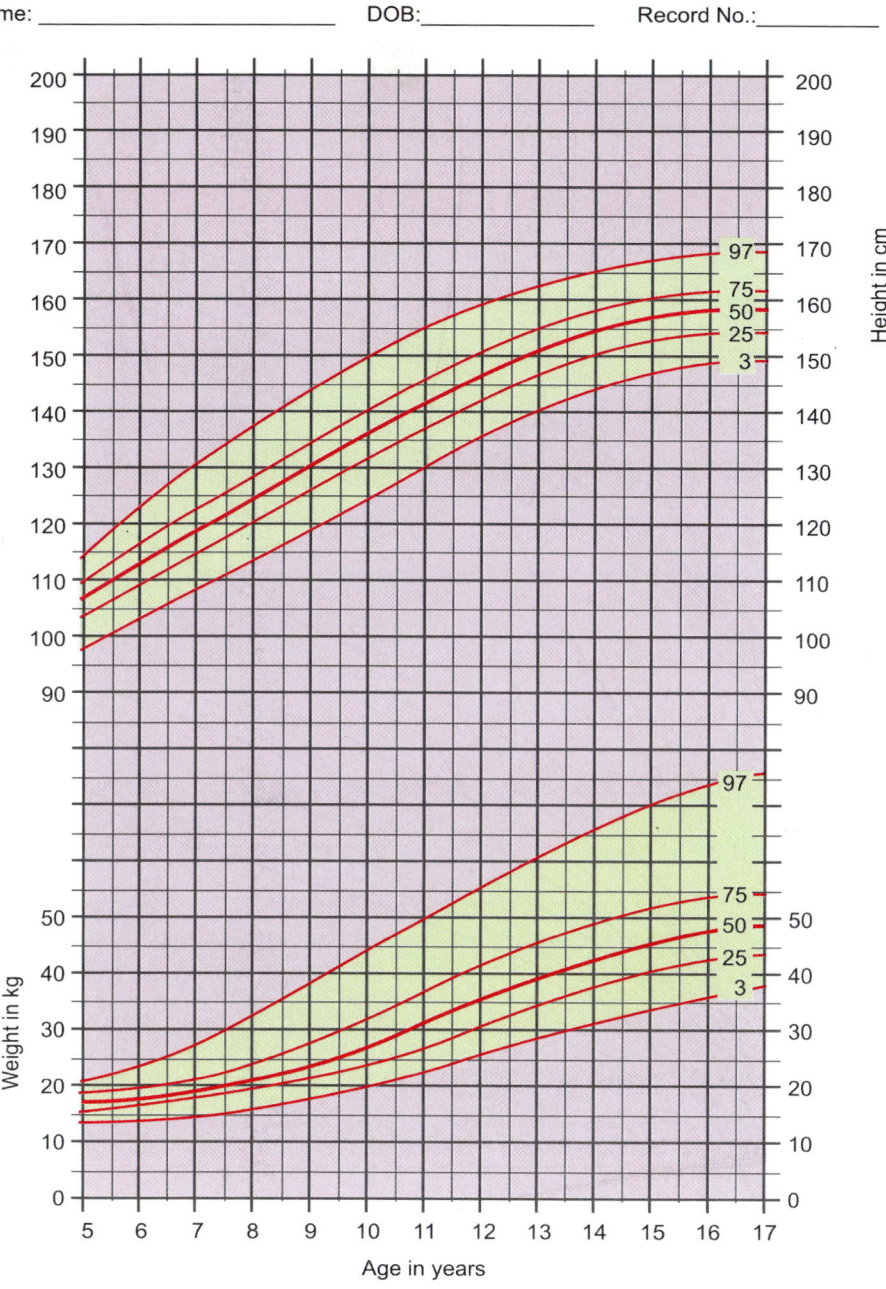

Height in cm

Weight in kg

Age in years

Adapted from: Agrawal DK, et al[13]

Fig. 1.10: *Growth chart for Indian children weight-for-age and height-for-age percentiles (Girls 5–17 years)*

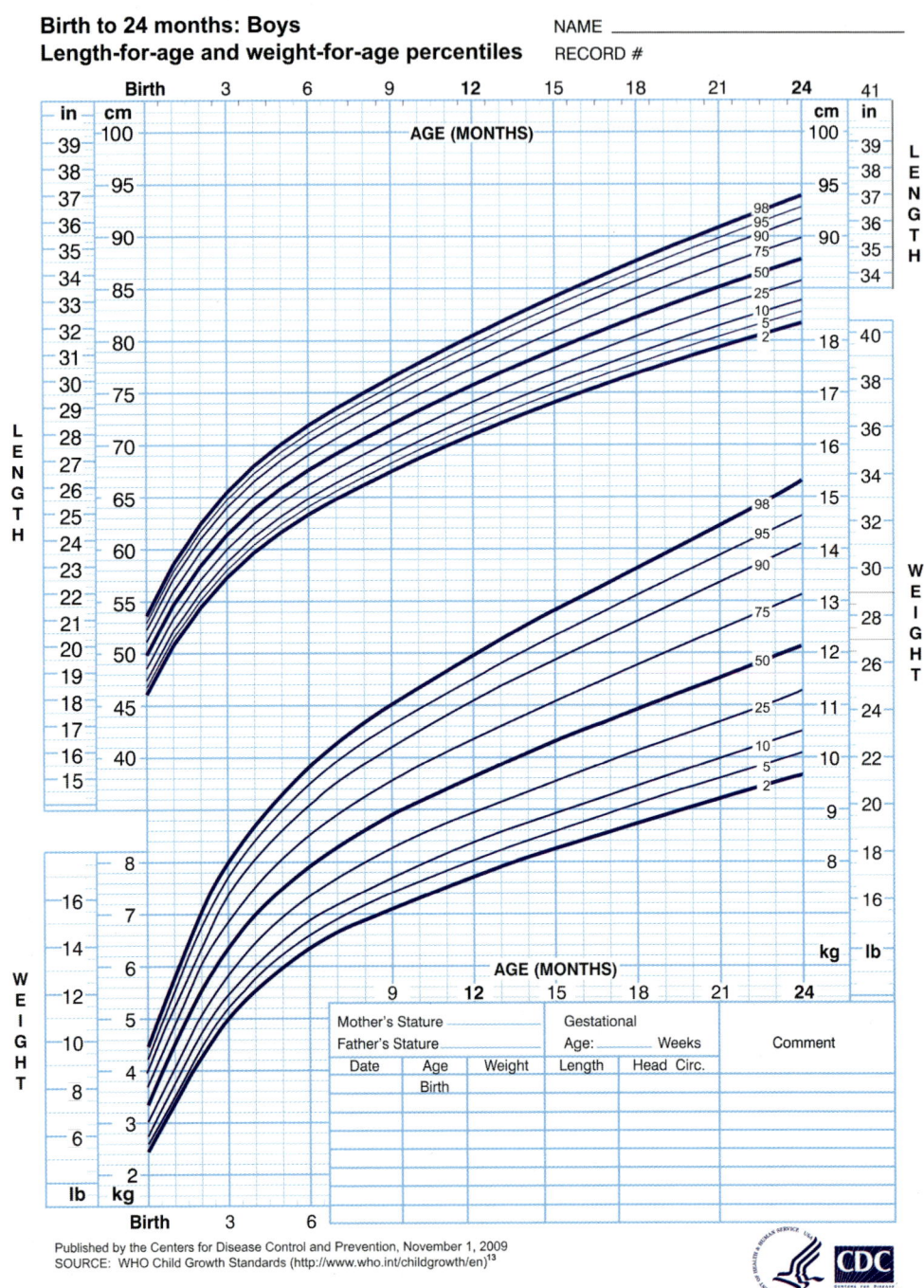

Birth to 24 months: Boys
Length-for-age and weight-for-age percentiles

NAME _____

RECORD # _____

Published by the Centers for Disease Control and Prevention, November 1, 2009
SOURCE: WHO Child Growth Standards (http://www.who.int/childgrowth/en)[13]

SAFER · HEALTHIER · PEOPLE™

Fig. 1.11: *WHO growth chart for boys*

Birth to 24 months: Girls
Length-for-age and weight-for-age percentiles

NAME _____

RECORD # _____

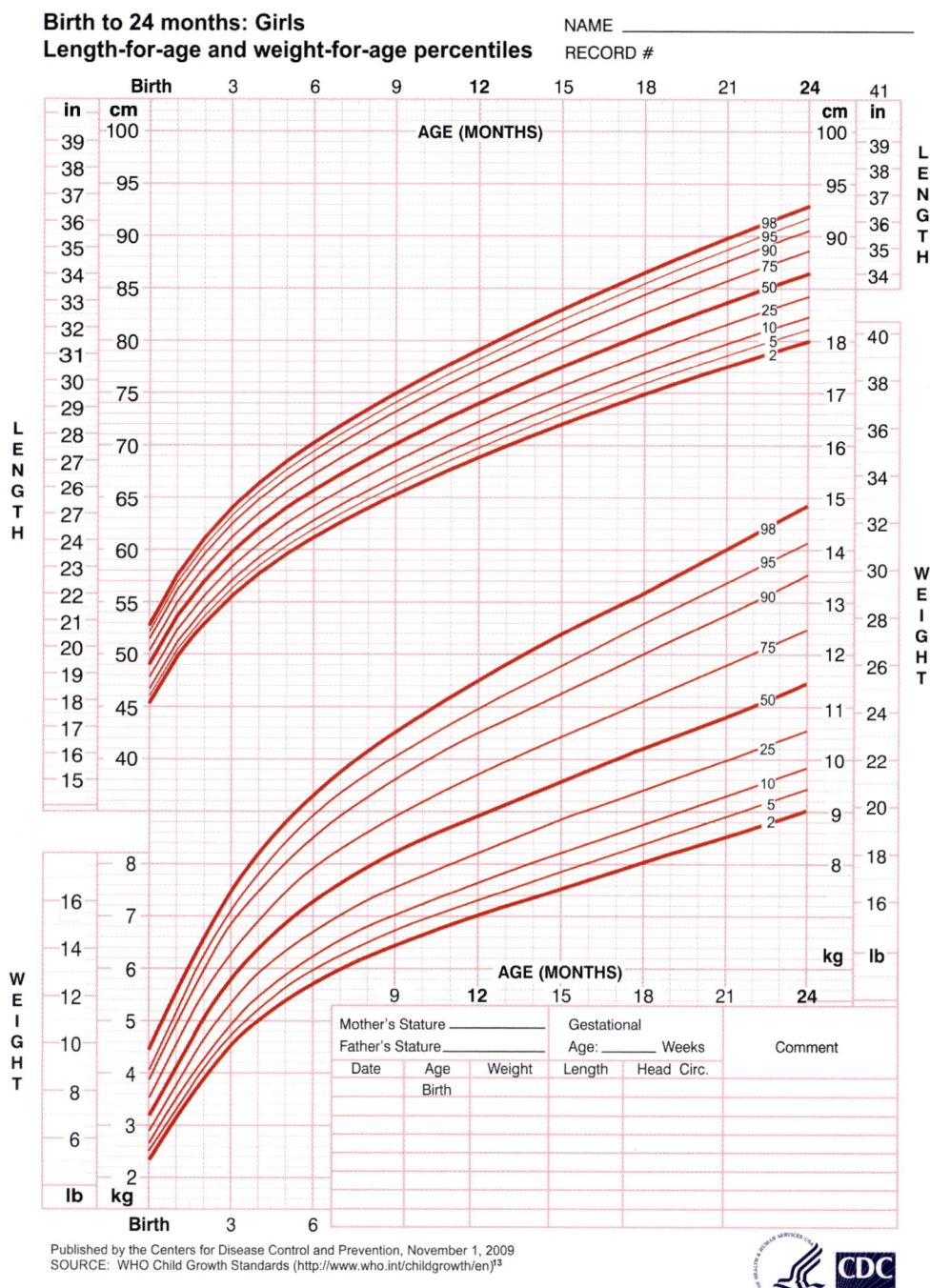

Published by the Centers for Disease Control and Prevention, November 1, 2009
SOURCE: WHO Child Growth Standards (http://www.who.int/childgrowth/en)[13]

Fig. 1.12: *WHO growth chart for girls*

above and below the target height 97th and 3rd centiles are constructed by tracing lines backwards to match the current age. If the projected and target height differ by more than 5 cm, then child need to be evaluated for pathological causes of short stature.

To summarize
- This height prediction is based on the sex adjusted midparental height:
  - *For girls:* Subtract 13 cm from the father's height and average with the mother's height.
  - *For boys:* Add 13 cm to the mother's height and average with the father's height.
- 13 cm is the average difference in height of women and men.
- For both girls and boys, 8.5 cm on either side of this calculated value (target height) represents the 3rd to 97th percentiles for anticipated adult height.
- Thus, a child's genetic potential for height is expressed by the mid-parental height (*see* page 6)

When to Worry in Growth Variations/Faltering?

(i) First three years
- Length/height, weight or head circumference below 3rd percentile or above 97th percentile on growth chart.
- Crossing of two major percentile lines (upward or downward), e.g. going from above 75th percentile to below 50th percentile on height or weight chart.
- A child below or above mid-parental range for height/length
- Weight loss or lack of weight gain for a month in the first 6 months.
- Absence of weight gain for 2–3 months from 6 to 12 months of age.
- Micropenis.
- Unilateral or bilateral undescended testis.
- Ambiguous genitals.

(ii) Three to nine years
- Length/height below 3rd percentile or above 97th percentile on growth chart.
- Crossing of two major percentile lines (upward or downward), e.g. going from above 75th percentile to below 50th percentile on height or weight chart.
- A child below by 5 cm mid-parental range for height.
- BMI over the 85th percentile at all ages.
- Rate of growth less than 5 cm/year.
- Girls with axillary, pubic hair growth or breast budding before 8 years and boys with axillary, pubic hair growth, genital growth or and testicular enlargement before 9 years.
- Children with craniospinal irradiation or surgery for brain tumors.
- Micropenis.

(iii) Nine to eighteen years
- Height below 3rd percentile or above 97th percentile on growth chart.
- Crossing of two major percentile lines (upward or downward), e.g. going from above 75th percentile to below 50th percentile on height or weight chart.
- A child below or above mid-parental range for height.
- BMI over the 85th percentile at all ages.
- Arrest at the same stage of puberty for more than 2 years.
- Micropenis.
- Unilateral or bilateral gynecomastia in boys.
- Hirsutism and menstrual irregularities in girls.
- Delayed puberty that is girls with no breast budding by 14 years or no menarche by 15 years and boys with no signs of puberty by 16 years.

Clinical Interpretation of Anthropometric Data

Please *see* Chapter 4, and Table 1.5.

Common causes of Growth Faltering

- Feeding difficulties, particularly in the younger child, i.e. failure of breastfeeding, unhygienic weaning foods, etc.
- Chronic ill health from whatever cause, including diarrhea, respiratory infection, malaria, tuberculosis.
- Social deprivation, where poverty and home circumstances are such that outcome is poor nutrition.
- 'Non-organic failure to thrive'.
- Child abuse.
- Overfeeding.

OTHER USEFUL BODY MEASUREMENTS FOR ASSESSMENT OF NUTRITIONAL STATUS

Mid-arm circumference (MAC): Measured mid-way between the acromion and olecranon processes (left upper arm), MAC increases from 9.8 cm at birth to 14.5 cm at 1 year of age. Between 1 and 5 years of age it shows a slow increase from 14.8 to 16.2 cm. Thus, for children in this age group MAC is used as a reliable tool to assess nutritional status. MAC of >13.5 cm is taken as normal, 12.5–13.4 cm is suggestive of mild to moderate undernutrition and <12.4 cm of severe undernutrition (Appendix Tables 16 and 17).

Chest circumference: For boys and prepubertal girls, chest circumference can be recorded at the level of nipple. In healthy children it exceeds head circumference by 8–10 months of age (Appendix Tables 18 and 19). In affluent Indian children chest circumference exceeded HC at 11.4 and 12.0 months, in boys and girls respectively.

Body mass index (BMI): The ratio of weight (in grams) to the square of height (in meters) (wt/ht^2), is a good indicator of variability in energy reserves. This is the single best indicator of nutritional status during adolescence. It is also an indicator of body fat and obesity (*see* Chapter 4, Tables 4.1 to 4.5) and Figs 1.13 and 1.14. Children with BMI >95th centile for their age and sex are considered to be obese, while those with BMI <5th centile are thin (*see* Chapter 4).

CDC–NCHS 2000 has recommended use of BMI in growth assessment from age 2 years onwards (Appendix Tables 5 and 6). The recent WHO 2006, growth curves are also available for BMI, birth to 5 years (Appendix Figs 1 and 2). BMI is gender and age specific for children and changes as children grow older. In adolescence, BMI height and weight are also dependent upon sexual maturity stages apart from age (Chapter 4, Tables 4.2 to 4.5). For adults however, BMI is neither age nor gender specific and nutritional status is defined by fixed cut-off points.

Ponderal index (PI): This is defined as the ratio of weight to the cube of height (wt/ht^3). Its main clinical application is in the newborn period where PI >2 indicates symmetrical IUGR (fetal growth and thinness, especially <6 years.

The various clinical interpretations from the anthropometric applications are given in Table 1.5.

Triceps and subscapular skin fold thickness (SFT): Skin fold thickness is an indicator of subcutaneous fat. It is measured over mid-point of the muscles by picking up a fold of skin and subcutaneous fat, by using *Lange's* or *Harpenden Calipers* (Appendix Tables 20 and 21).

Children having tricep and sub-scapular skin fold >95th centile are considered obese. The sum of SFT measured over triceps, biceps, subscapular and suprailiac regions is used to assess % of body fat in children (conversion table provided with ranges caliper M/s Cambridge Scientific Ind. Cambridge MD 21613). On the other hand, triceps and subscapular SFT <5th centile indicates of thinness (Appendix Tables 20 and 21).

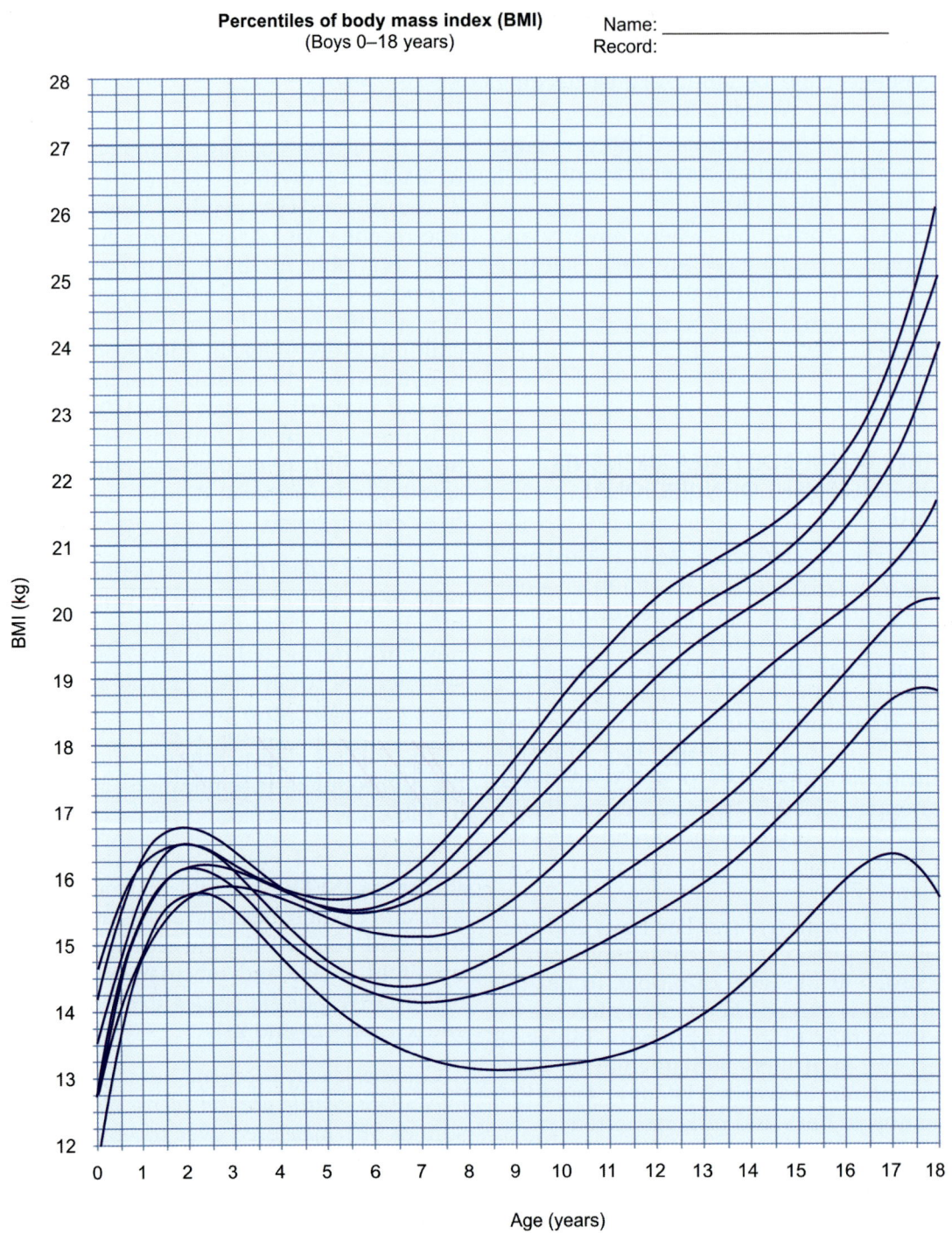

Percentiles of body mass index (BMI)
(Boys 0–18 years)

Name: _____
Record:

Age (years)

Fig. 1.13: *Percentiles for body mass index (BMI) for boys birth to 18 years) of age (Indian affluents*
Source: Agarwal[9, 13]

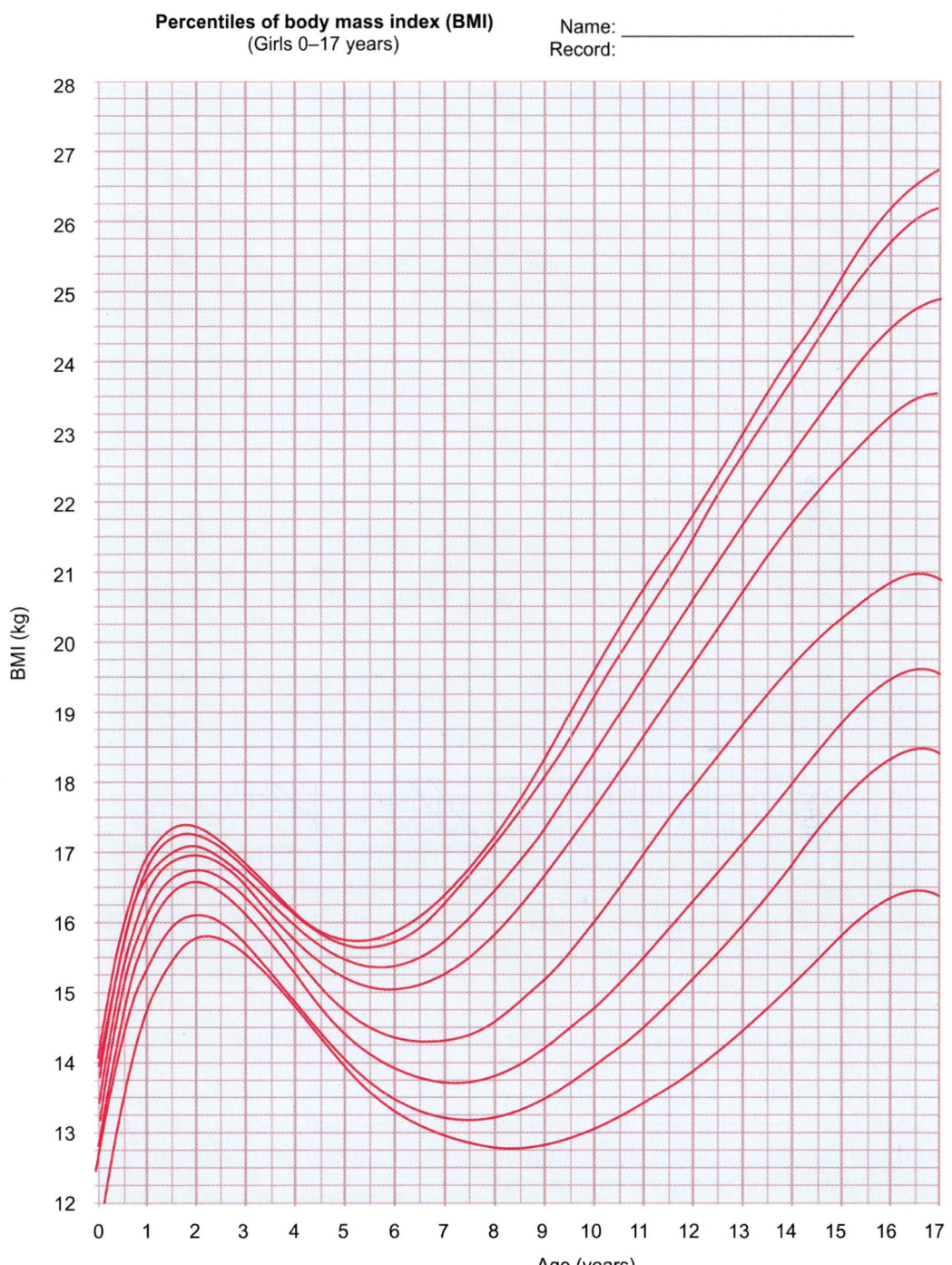

Percentiles of body mass index (BMI)
(Girls 0–17 years)

Name: _____
Record:

Fig. 1.14: *Percentiles for body mass index (BMI) for girls birth to 17 years of age (Indian affluents Source: Agarwal[9, 13])*

Table 1.5: *Growth measurement in clinical applications*

Anthropometric indicator	Terms describing outcome	Terms describing process	Explanation
Low height (ht.) for age (<3rd centile)	**Shortness**	Gaining insufficient height relative to age	Possibly genetic in normal population. Familial, long-term malnutrition chronic systemic disease endocrine disorders
	Stunted (<90% of reference; or <–2 Z-scores*)		
Low weight for height (wt./ht.)	Thinness		
	Wasted (<80% of reference wt./ht)	Gaining insufficient weight relative to height	Implies recent or continuing current weight loss
Low weight-for-age	—	—	Not necessarily pathological implies wasting if also associated with low weight for height
	Underweight (<80% of reference or <–2 Z-score)*	Gaining insufficient weight relative to age or losing weight	
High weight-for-height	Overweight BMI >85th centile	Weight more than the ref. weight for height	Could be nutritional or 'simple' or due to underlying systemic illness 'pathological'

$$* \text{Z-score} = \frac{\text{Observed value} - \text{Median reference}}{\text{SD of reference population}}$$

Growth Studies in India: Analytical Review

In 1971, a committee of International Union of Nutrition Sciences[18] made recommendations for establishment of growth standards. They stated that "The commission strongly recommends that studies be carried out in as large a variety of countries as possible. Each country's own standards must be derived from carefully selected samples representing children growing in an optimal environment for that country." In selecting the specific population the first group should be from the "Modern elite group in each study area". The argument for use of "Growth standards" based on elite population is due to the fact that this group has no nutrition constraints and also has all possible access to the best possible health care. Secondly, it is also justified as all individuals in the community have a right to attain status of growth and health enjoyed by the privileged section of the society.

The justification for the creation of growth standard is as follows:

1. Anthropometric measures are the most important indices of assessing growth and nutritional status.
2. Appropriately developed standards can serve as a reference against which to measure changes in health and nutrition of a given country—secular trends.
3. Can be used as standards for evaluating the results of intervention programs.

Growth Studies (Agarwal et al)[9,13,14]

The data on affluent Indian children were collected during 1989–1991 (Nutrition Foundation of India—NFI) from birth to 5 years (7 states), only full term with birth weight of 2500 g and above (boys 433 and girls 346) were followed during first year of life at 3, 6, 9 and 12 months of age with minimum of 3 reading for every infant (cohort I). In cohort II, from 12 months to 5 years (children of cohort I also continued) of age 1011 boys and 874 girls were followed on their birthday and 6 monthly with minimum of 3 measurements for each child up to 72 months of age. Children had received exclusive breast milk for 3–4 months of life in cohort I and II (as prevalent in those years).

The Indian Council of Medical Research (ICMR) cross sectional multicenter data for physical growth and sexual development for 5 to 17 years in girls and from 5 to 18 years in boys (9 states, 23 schools; 12893 boys and 10,941 girls) on affluent Indian children were collected[13, 14] during 1989–1991. These two data sets[9, 13] are collected around same time on affluent Indian children 'Birth to adolescence'. These data sets on physical growth and sexual development—birth to 17 and 18 years of age for girls and boys respectively will continue to serve as the baseline **reference** (Natale and Rajgopalan)[15] data for assessing physical growth, and sexual development—assessed by the same measurers in all centers in the country. These data remain the only 'Sexual character study' with growth in Indian school children till date. (*see* Tables 22 to 25 in Appendix).

The WHO growth data,[12] in India were collected from south Delhi area, and pooled in the International data. The WHO Multicenter Growth Reference Study (MGRS) was undertaken between 1997 and 2003 to generate new growth curves for assessing the growth of breastfed infants and young children around the world. The MGRS collected primary growth data and related information from approximately 8500 children from widely different ethnic backgrounds and cultural settings (Brazil, Ghana, India, Norway, Oman and USA). These growth curves are expected to provide a **single international standard** that represents the best description of physiological growth for all children from birth to five years of age and to establish the breastfed infant as the normative model for growth and development. These growth charts depart from the growth *reference* model in several ways. Children from six countries provided the data measurements, which were not representative of their country of residence, and were selected on the basis of *socio-demographic* criteria and child's nutrition as per WHO guidelines. In preparing WHO growth charts children with higher weight than is considered compatible are excluded. Further to make these data comparable several statistical adjustments had to be made. In the USA, these standards are recommended to monitor growth for infants and children from birth to 2 years of age.

Comparison of Agarwal Growth Data[9, 13, 14] with the WHO (MGRS) Standards[12]

In a recent study "Worldwide variation in human growth and the WHO growth standards: A systemic review" by Natale and Rajgopalan.[15]

WHO assumed that all economically advantage children who were breastfed as infants grow similarly? As a result, a single set of growth charts can be used to judge growth in any child regardless of race or ethnicity. Natale and Rajgopalan[15] compared mean heights, weights and head circumferences from a variety of studies with the WHO's data. They compared WHO (Multicenter Growth Reference Study—MGRS)[12] with data from 55 countries or ethnic groups. These study countries included India (data from Agarwal and Agarwal 1994),[9] Norway and the USA as these countries had participated in the MGRS. For height, Indian children were below 0.5 SD at more than 3 age points (birth to 5 years of age). In a large German study 2011 by Rosario et al[19] found that means for girls and boys were taller being at 62nd and 60th MGRS centiles respectively, this made them to use the national data as standards instead of MGRS. Weight and head circumference varied more than the height.

The study concluded that height and weight may not be the optimal fits in all cases; similarly **MGRS head circumference means may put many children at risk of misdiagnosis of micro/macrocephaly**. The study recommended that country reference data may be more ideal for assessing growth.

Advantages of WHO Charts

Birth to 5 years: Purely breastfed followed by emphasis on breastfed.

Disadvantages of WHO 2007 (Birth to 18 years) Charts

1. It assumes all populations in the world are equally tall. After 5 years, it extends the lines to 18 years statistically based on NCHS charts. The fact is that Indian affluent children of as recent as Khadilkar et al[20] data do not have the same pubertal growth spurt as do Americans, and our final height is shorter.
2. BMI centiles at 18 years are heavier than prescribed BMI for adult Indian cut-off.

What have Other Countries done?

1. *USA:* Felt from 1977 (1 year after 1976 NCHS charts were made) that there were some inaccuracies in the NCHS charts, mainly pertaining to head circumference data of birth to 5 years, as well as preterm data, and statistical smoothening procedures. They convened many workshops, which resulted in the CDC 2000 charts. While making the CDC charts, they **excluded weight data of NHANES III**, on the grounds that the children had become too heavy and that misclassification of obesity and overweight would occur.
2. *UK:* Felt on account of breastfed, and preterm representation, should adopt WHO standards and they made UK–WHO 2009 charts in May 2009. This is **only for birth to 4 years**. For 4 to 18 years,

UK 90 charts are still continued to be used.

In recent years two more studies have appeared one being multicenter by Khadilkar et al[20] and second from Delhi schools by Marwah et al.[21] The distribution of study subjects is shown below with comments.

Khadilkar et al[20] data is more skewed towards right side. It has more extreme values on the right of the median as compared to Agarwal et al[9, 13] data. Figure shows that Khadilkar et al data have more dispersion in comparison to Agarwal et al[9, 13] data (Figs 1.15 a to d).

The density curves comparing primary data dispersion of Agarwal et al[9, 13] and Marwah et al[21] data are closer (Fig. 1.16).

Khadilkar et al[20] data line for 3rd centile flattens after 14 years of age (skewing); as compared to Agarwal et al[9, 13] and Marwah et al[21] (Fig. 1.17).

1. Further comparing Agarwal et al[9, 13] school data (ICMR) against recently available Khadilkar et al[20] data are important in that they take a look at secular trend in height and weight over the last 20 years among children from affluent families having no constraints to nutrition and health. They found no increase in final height at 18 years at the 3rd centile; only 0.6 cm increase in boys at 50th centile, and 1.7 cm (in boys) to 2 cm (in girls) increase in height at the 97th centile.
2. For weight, the 50th centile boys were 2.9 kg heavier at 18 years and 97th centile were 14.7 kg heavier. For girls, the respective figures were 8.0 kg at 17 years.

For BMI, Khadilkar et al[20] 85th and 95th centiles were 26.2 and 30.3 kg/m^2 in boys and 25.9 and 29.9 kg/m^2 in girls. Agarwal et al[13] 85th and 95th for boys were 23.6 and 28, and for girls 23.0 and 25.9 at 17 years. The adult cut-offs recommended for overweight and obesity in Indians are 23 and 28 kg/m^2.

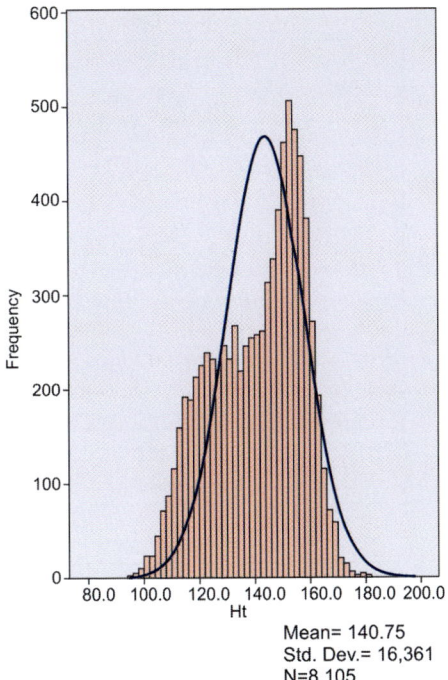

Mean= 140.75
Std. Dev.= 16,361
N=8,105

Fig. 1.15a: *Distribution curve—Weight in girls (Khadilkar et al)*

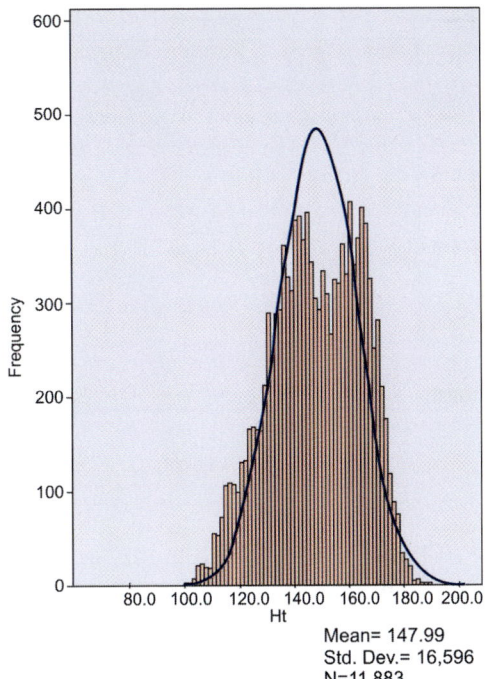

Mean= 147.99
Std. Dev.= 16,596
N=11,883

Fig. 1.15b. *Distribution curve—Weight in girls (Agarwal et al)*

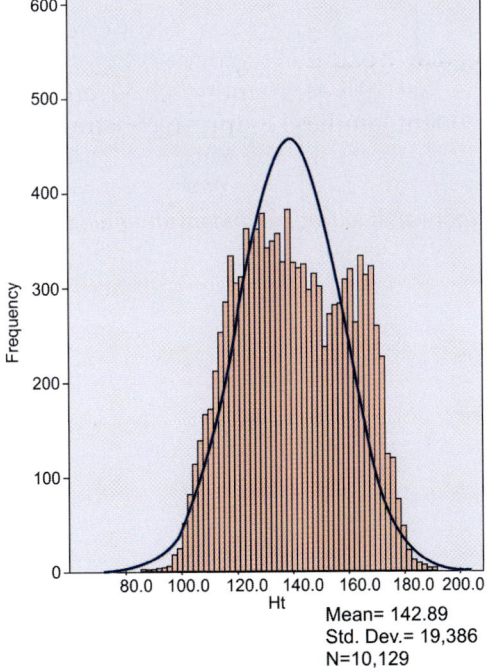

Mean= 142.89
Std. Dev.= 19,386
N=10,129

Fig. 1.15c: *Distribution curve—Weight in boys (Khadilkar et al)*

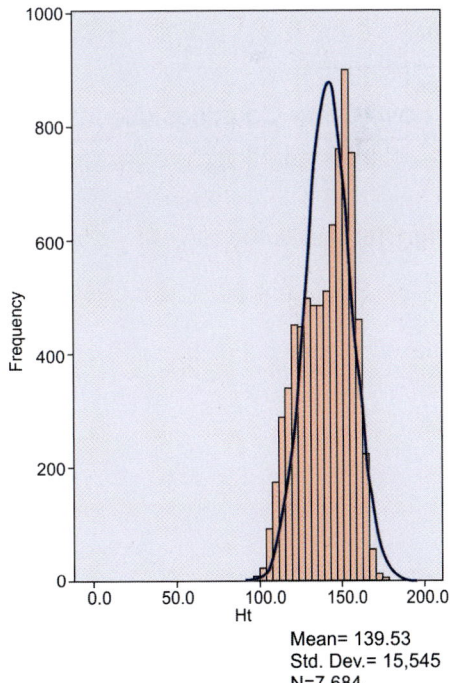

Mean= 139.53
Std. Dev.= 15,545
N=7,684

Fig. 1.15d: *Distribution curve—Weight in boys (Agarwal et al)*

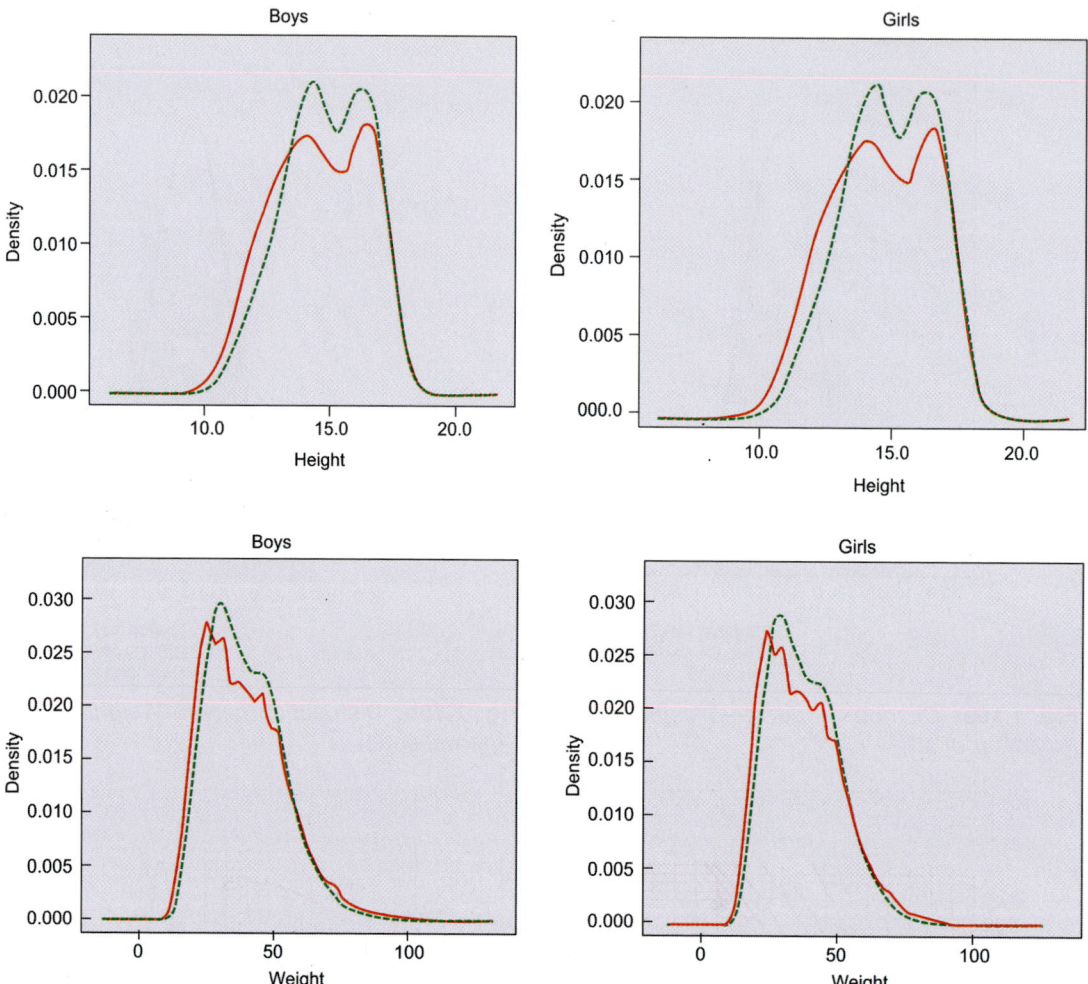

Fig.1.16: *Distribution curves for primary data from Agarwal et al (red) and Marwah et al (green)*

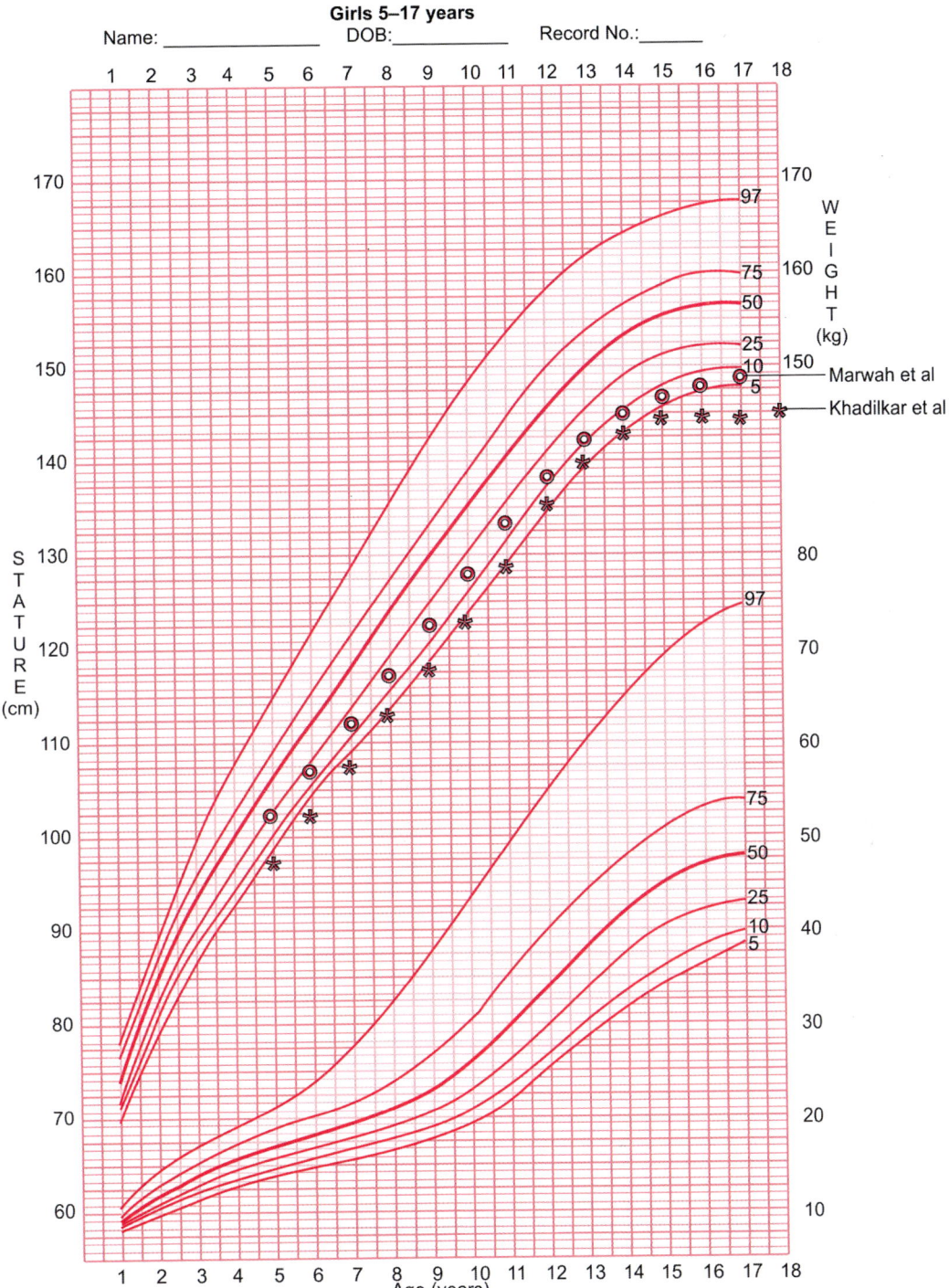

Fig. 1.17: On Agarwal[9, 13] female growth curve plotting of 3rd centile of Khadilkar et al[20] showed drop around 14 years of age. Dots from Marwah et al[21] follow the centile line. (Courtesy: Prof Vijayalakshmi Bhatia, SGPGI, Lucknow)

Comparing Growth Velocities—Height/Weight in 3 Data Sets

Comparison of Girls Height

Peak of height velocity in girls in 1990 data was between 10 and 11 years, that for the two contemporary data sets is between 9 and 10 years.

Peak of height velocity in boys in 1990 data was between 13 and 14 years, that for the two contemporary data sets is between 12 and 13 years.

Comparison of Girls Weight

Peak of weight velocity in girls in 1990 data was between 11 and 12 years that for the two contemporary data sets is between 10 and 11 years.

Peak of weight velocity in boys in 1990 data was between 14 and 15 years, while that for the two contemporary data sets is between 12 and 13 years.

Comparative growth velocities in three data sets under discussion are shown in Fig. 1.18. The earlier onset of peak height velocity in studies by Khadilkar et al[20] and Marwah et al,[21] is a known effect of obesity.

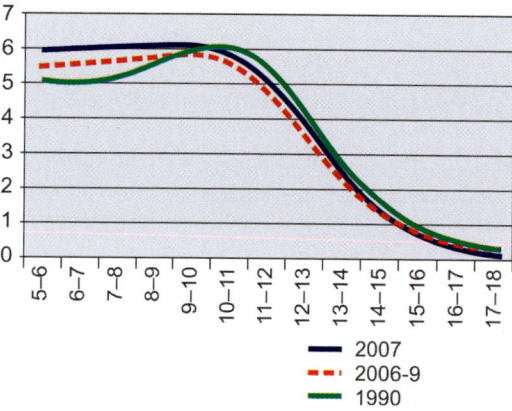

Fig. 1.18a: *Showing peak height velocity (girls) 1990: Green (Agarwal), 2006–09: Red (Marwah), 2007: Blue (Khadilkar)*

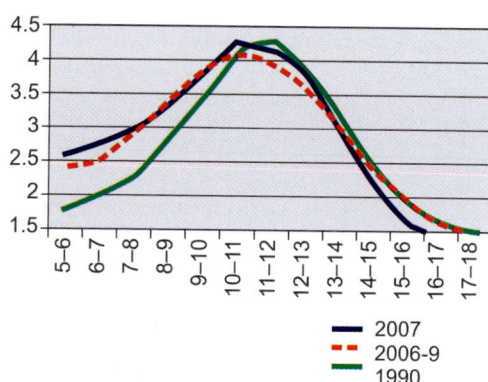

Fig. 1.18c: *Peak weight velocity in girls*

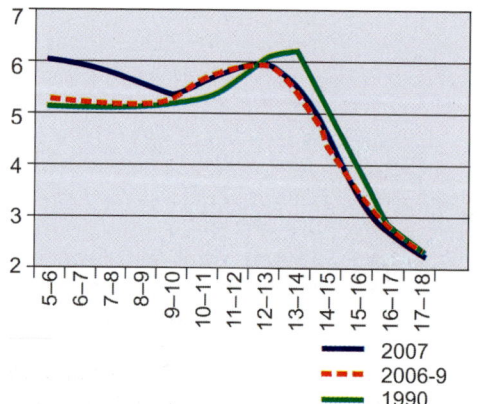

Fig. 1.18b: *Peak height velocity in boys*

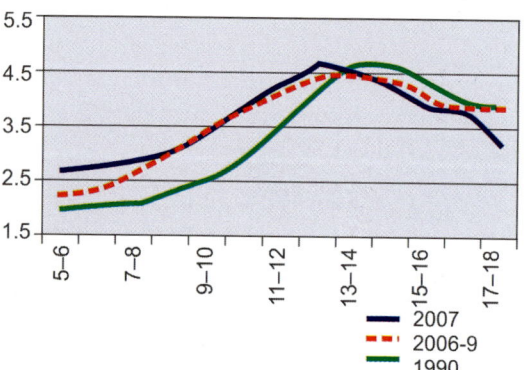

Fig. 1.18d: *Peak weight velocity in boys*

Choice of Growth Curves

Redoing the curves in response to increasing weight alone will have the effect of 'normalizing' the weight gains you are seeing at the higher end of the scale since the definition of 'normal' weight will shift to the right, potentially putting more children who today are classified as overweight into the normal weight category. At the bottom of the scale, some children who are today classified as normal weight will be classified as underweight. On both the high and low ends of the scale, creating new reference curves with these changes may not be beneficial from a public health perspective (*Ronald E. Kleinman, Professor and Head Pediatrics, Harvard Medical School, Boston*).

It is important to take lesson from the "Methods including the data source in construction of the CDC 2000 growth charts", as weight data from the National Health and Nutrition Examination Survey (NHANES) III (1988–94) were excluded from the weight-for-age and BMI-for- age curves because of a secular trend in body weight that occurred between NHANES II (1976–80) and NHANES III.[3]

Presently, the choice remains using Agarwals, growth charts based on data (birth-adolescence), this was also recommended in growth chart evaluation study by Khadgawt at et al.[22] The WHO curves may be used up to 2 years of age[12] as practiced in the USA, however, may not be ideal as comparative review study by Natale and Rajgopalan[15] recommend to use the Indian growth data as geographical and ethnic variations are significant. The continued use of WHO growth standard will assess more underweight and short statured Indian children. Further, the Indian affluent children at 18 years of age are still shorter than the NCHS/WHO data. Thus we must examine our children in nationally collected growth data sets; as calculated BMI values are also different as compared to the WHO/NCHS values Agarwal et al[9, 13] (BMI—

Tables 4.2–4.5 and NCHS appendix Tables 5 and 6). From 9 years in girls and 11 years in boys sexual assessment is important to assess physical growth for which child can be provided the Tanner's sexual maturity rating diagrams[11] and he/she (breast and pubic hair) simply write the Stage 1–5 (*see* Figs 2.1 and 2.2). For boys Prader's orchidometer will be of use to measure privately the testicular volume (Fig. 2.3).

REFERENCES

1. Doublet PM, Benson CB, Nadel AS, et al. Improved birth weight table for neonates developed from gestations dated by early ultrasonography. Journal of Ultrasound Medicine 1997;16:241.

2. Hadlock FP, Shah YP, Kanon DJ, et al. Fetal crown rump length: Re-evaluation of relation to menstrual age with high resolution real-time US Radiology 1992;182:501.

3. Ghosh S, Bhargava SK, Moriyamma IM. Longitudinal study of survival and outcome of a birth cohort, Vol II. Department of Pediatrics, Safdarjung Hospital, New Delhi India, 1979.

4. Mittal SK, Singh PA, Gupta RC. Intrauterine growth and low birth weight criteria in Punjab infants. Indian Pediatr 1976;13:679–82.

5. Bhatia BD, Bhargava V, Chatterjee M, et al. Studis on fetal growth patterns: Intrauterine growth percentiles for singleton live born babies. IBID 1981;18:647–53.

6. Agarwal KN, Agarwal DK, Agarwal A, et al. Impact of the Integrated Child Development Services (ICDS) on maternal nutrition and birth weight is rural Varanasi. IBID 2000; 37:1321–27.

7. Agarwal S, Agarwal A, Bansal AK, et al. Birth weight pattern in rural undernourished pregnant women. IBID 2002;39:244–53.

8. Lubchenko LO, Hausman C, Boyd E. Intra-uterine growth in length and head circumference as estimated from live births at gestational ages from 26 to 42 weeks. Pediatrics 1966;37:403–8.

9. Agarwal DK, Agarwal KN. Physical growth in affluent Indian children (birth to 6 years). Indian Pediatr 1994;31:377– 413.

10. Tanner JM, Whitehouse RH, Takaishi M. Standards from birth to maturity for height, weight, height velocity and weight velocity in British Children Arch. Dis Child 1965; 41:454–71 and 613–35.

11. Tanner JM, Whitehouse RH. Clinical longitudinal standards for height, weight and the stages of puberty. Arch Dis child 1976;51:170–9.

12. WHO Multicentric Growth Reference Study Group: WHO child growth standards based on length/height, weight and age. Acta Pediatr 2006; Suppl 450:76–85.

13. Agarwal KN, Agarwal DK, Upadhaya SK, et al. Physical and sexual growth pattern of affluent Indian children from 5 to 18 years of age. Indian Pediatr 1992;92:1203–82.

14. Agarwal KN, Saxena A, Bansal AK, et al. Physical growth assessment in adolescence. IBID 2001;38:1217–35.

15. Natale V, Rajgopalan A. Worldwide variations in human growth and the World health Organization growth standards: A systematic review. BMJ Open 2014;4:1–11.

16. Singhal A, Lucas A. Early origins of cardiovascular disease: Is there a unifying hypothesis 2004;363:1642–5.

17. Buyken AE, Karopis-Danckert N, Remer T, et al. Effects of Breastfeeding on trajectories of body fat and BMI throughout childhood. Obesity 2008;16:389–95.

18. International Union of Nutrition Sciences. The creation of growth standards. A committee report of meeting in Tunis. Am J Clin Nutr 1971;25:218–20.

19. Rosario AS, Schienkiewitz A, Neuhauser H. German height references for children aged 0 to 18 years compared to WHO and CDC growth charts. Ann Hum Biol 2011;38:121–30.

20. Khadilkar VV, Khadilkar AV, Cole TJ, et al. Cross sectional growth curves for height, weight, and body mass index for affluent Indian children, 2007. Indian Pediatr 2009;46:477–85.

21. Marwah RK, Tandon N, Singh Y, et al. A study of growth parameters and prevention of overweight and obesity in school children of Delhi. Indian pediatr 2006;43:943–52.

22. Khadgawat R, Dabadghao P, Mehrotra RN, Bhatia V. Growth charts for evaluation of Indian children. Indian Pediatr 1998; 35:859–65.

- **Agarwal et al 1992 height, weight, data reanalyzed using LMS method—Graphs and Tables are given in Chapter 11; pages 105.**
- **The 2015 IAP growth chart are available on the IAP website. These are constructed on studies published by different authors. I have examined these curves it appears that the obesity effect has greatly influenced these growth charts, remain *unsuitable to detect early overweight and obesity* and at the same time over estimate undernutrition.**

2

The Somatic Growth and Sexual Development in Adolescence

DK Agarwal

DEFINITION

World Health Organization (WHO) identifies adolescence as the period in human growth and development that occurs after childhood and before adulthood, from ages 10 to 19 years. It represents one of the critical transitions in the lifespan and is characterized by a tremendous pace in growth and change that is second only to that of infancy. Biological processes drive many aspects of this growth and development, with the onset of puberty marking the passage from childhood to adolescence. The biological determinants of adolescence are fairly universal; however, the duration and defining characteristics of this period may vary across time, cultures, and socio-economic situations. This period has seen many changes over the past century namely the earlier onset of puberty, later age of marriage, urbanization, global communication, and changing sexual attitudes and behaviors.

ADOLESCENT GROWTH AND DEVELOPMENT

Adolescence is the time between the beginning of sexual maturation (also known as puberty, from the Latin pubertas, meaning adult) and the beginning of adulthood.

Adolescence usually spans the years between ages 10 to 16 in girls and 12 to 19 in boys. Adolescence includes:

a. Physical growth,
b. Sexual development,
c. Emotional,
d. Psychological, and
e. Mental change

Adolescence is a period of many transitions. During the teen years, adolescents experience changes in their physical development at a rate of speed unparalleled since infancy. Physical development includes rapid gains in height and weight. During a one-year growth spurt, boys and girls can gain an average 10–11 cm and 8–9 cm in height, respectively. This spurt typically occurs two years earlier for girls than for boys. Weight gain results from increased muscle development in boys and body fat in girls. These changes are accompanied by the development of primary sex characteristics including the further maturing of the gonads, the testes in boys, and the ovaries in girls. In response, the gonads produce a variety of hormones that stimulate the growth, function, or transformation of brain, bones, muscle, blood, skin, hair, breasts, and sex organs.

NEUROENDOCRINAL CONTROL OF PUBERTY

Hormones play two different roles in adolescent development, the organizational role and the activational role. The organizational role is the ability of hormones to generate different patterns of behavior in the male and in the female brain, respectively. Already in the prenatal age, the hormones organize the brain differently. The activational role is the ability to initiate the modifications related to puberty and to differentiate them for male and females. During and just before puberty, the hypothalamus both stops the inhibition upon the factors able to initiate puberty and begins to produce substances that set the puberty in motion. The first signal begins due to higher concentrations of leptin, a protein produced by adipocytes of the fat tissue. The hypothalamus stimulates the hypophysis to secrete hormones able to promote the overall growth of the body and to mature gonads, adrenal cortex and thyroid. Adrenal cortex maturation is involved in sexual attraction. Hormone concentrations are due to gland activations that are controlled by several mechanisms of feedback.

In both sexes, hormonal regulation of reproduction is regulated by the brain. Until eight weeks of gestation, the brain is organized in a female direction irrespectively with the gender of the fetus. Successively, testosterone, for example, organizes the male brain in patterns of behavior, many of which may not appear until much later. Hypothalamic gonadotropin releasing hormone (GnRH) controls release of both luteinizing hormone (LH) and follicle stimulating hormone (FSH). LH acts primarily on endocrine cells of the gonads. FSH acts primarily on gamete-producing cells. Both sexes produce androgens and estrogens. Testosterone, the main androgen produced in the testis is converted to dihydrotestosterone (DHT) in many tissues. Estradiol, the main estrogen, is made from testosterone by the action of the enzyme aromatase. Both the ovary and testis produce peptide hormones that have feedback effects on the hypophysis. Inhibins are hormones that inhibit FSH secretion. Activins stimulate FSH secretion as well as spermatogenesis, oocyte maturation. Children (both males and females) with a deficiency of GnRH will not mature in the absence of gonadotropin stimulation due to lower levels of androgens and estrogens.

During puberty, changing hormonal levels play a role in activating the development of secondary sex characteristics. These include:

i. Growth of pubic hair (pubarche);
ii. Growth of the breasts in girls (thelarche)
iii. Menarche (first menstrual period for girls) or penis growth (for boys);
iv. Voice changes (for boys);
v. Growth of under arm hair;
vi. Facial hair growth (for boys);
vii. Night time ejaculations (nocturnal emissions; 'wet dreams' for boys, and
viii. Increased production of oil, increased sweat gland activity, and the beginning of acne.

ADOLESCENT DEVELOPMENT

Growth in Adolescence as per Centers for Disease Control (CDC)

The following are some average ranges of weight and height, based on growth charts developed by the Centers for Disease Control and Prevention (CDC): Briefly, at 12 years of age a male, should be 54–63.5 in. (1.37–1.6 m) and a female 55–64 in. (1.4–1.6 m). The weight would be 66–130 lb. (29.9–58.9 kg) and 68–136 lb. (30.8–61.7 kg) respectively. At 18 years of age, a male would be 65–74 in. (1.7–1.9 m) tall and a female 60–68.5 in. (1.5–1.7 m). The weight would be 116–202 lb. (52.6–91.6 kg) and 100–178 lb. (45.4–80.7 kg), respectively.

Growth not only involves length and weight of a body, but also includes internal growth of every organ and development, including the brain. Growth also affects different parts of the body at different rates:

i. The head reaches almost its entire size by age one;

ii. Throughout childhood, a child's body becomes more proportional to other parts of his/her body.

iii. Growth is complete between the ages of 16 and 18 years, at which time the growing ends of bones fuse. Although a child may be growing, his/her growth pattern may deviate from the normal. Ultimately, the child should grow to normal height by adulthood. Teens frequently sleep longer.

iv. Research suggests that teens actually need more sleep to allow their bodies to conduct the internal work required for such rapid growth. On average, teens need about 9.5 hours of sleep a night.

Normal Female Development

Sexual development in girls; breast enlargement (thelarche, B_2), occasionally initially unilateral, is the first obvious sign of puberty and occurs between 10 and 11 years of age. The Peak Height Velocity (PHV), initiates at this stage of thelarche and maximum height gain occurs during the period of sexual maturity breast stages B_2 to B_3. Gradually the breast diameter increases and the areola darkens and becomes more prominent. The approach to good nutrition and health care will achieve maximum physical growth, sexual development and mental functions (Appendix Table 24).

Pubic and axillary hair growth in girls is a sign of adrenal androgen secretion. It starts at about the time of apocrine gland sweat production and the common complaint of axillary odour.

Menarche usually occurs about 2–3 years after the start of breast development B_2

(thelarche). The median age of menarche is around 12 year 6 months in Indian girls (Appendix Table 25). The median age of menarche is around 13 years in contemporary British teenagers (12 years 11 months). The growth spurt occurs early in female puberty, it is usually maximal (about 8 cm/year) during Tanner breast stages 2 and 3 (B_2 and B_3). But reduces to 4 cm/year at menarche. In the postmenarcheal period maximum height gain remains around 5 cm only. Thus, the maximum height gain is in pre-menarche in SMR—stages B_2 and B_3. Puberty leads to the enlargement of the sex organs and external genitalia as well as thickening of the endometrium and vaginal mucosa.

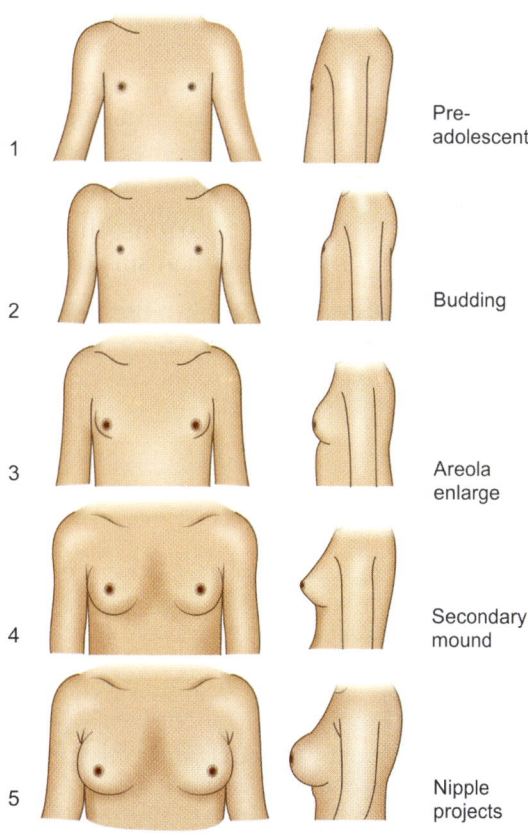

1   Pre-adolescent

2   Budding

3   Areola enlarge

4   Secondary mound

5   Nipple projects

Fig. 2.1: *Breast development (B 1–5) according to Tanner 1962*

SUMMARY

Puberty–Girls

1. First sign of ovarian estradiol secretion is breast development "Thelarche". SMR-B$_2$ (breast budding)—Growth in height.
2. Estradiol is a good stimulator of "GH" it doubles the growth velocity 'peak height velocity' (9–10 cm/yr). Coincides with B$_3$ follows B$_2$ by 1 year.
3. Change in body shape
4. Growth under arm hair followed by secretion
5. Menarche follows PHV by 14–18 months
6. Adult size breast

Normal Male Development

Sexual development in boys: Testicular enlargement usually occurs between ages of 12 and 13 years. The pre-pubertal testes are about 2 ml in volume, with puberty taken to begin when a volume of around 4 ml is attained. Testicular growth starts as early as 10 years of age, associated with enlargement of seminiferous tubules, epididymis, seminal vesicles and prostate.

Penile and scrotal enlargement occurs typically about a year after testicular enlargement. Pubic hair typically appear at a similar time (Appendix Tables 22 and 23, Indian affluent boys).

The growth spurt occurs later than in girls, possibly because testosterone is less of a stimulus to growth hormone responsiveness than estradiol in girls and is required in relatively higher concentrations of testosterone to produce the same anabolic effect. A greater and later growth spurt occurs in boys and ultimately achieves an average 12–13 cm greater height in adult men than the female counterparts. The growth spurt is on an average 2 year later than girls thus boys get 2 year extra height gain of 10–12 cm (5–6 cm per year). Although boys are on average 2 cm shorter than girls before puberty begins, finally adult men are on average about 13 cm (5.2 inches) taller than women. Most of this sex difference in adult heights is attributable to a later onset of the growth spurt and a slower progression to completion, a direct result of the later rise and lower adult male levels of estradiol.

SUMMARY

Puberty—Boys

1. Adrenarche is the onset and continuity of male puberty
2. Testosterone/dihydrotestosterone are needed in large concentration to initiate "GH" via the androgen receptors. (Thus, later than girls by 1–2 year)
3. Initiation testicular volume >4 ml; maximum growth "PHV" (10–11 cm/year) attained at testicular volume 10–12 ml (during SMR-G3–4)
4. Testosterone—deepens the voice and increases body muscle mass (lean body mass)

Table 2.1: *Development of breast and pubic hair in girls (see fig. 2.1)*

SMR	Breast (mean ages)*	Pubic hair
1.	Pre-adolescent	Pre-adolescent
2.	Bud stage and papilla elevated as small mound, areolar diameter increased (10.2 yr)	Sparse lightly pigmented straight, medial border of labia (in 22%)
3.	Areola enlarged (11.6 yr) no contour separation	Darker, beginning to curl, increase in amount (in 92%)
4.	Areola and papilla form secondary mound (13.5 yr)	Coarse curly, abundant but amount less than in adult (98%)
5.	Mature nipple projects, areola part of general breast contour (15.6 yr)	Adult feminine triangle spread to medial surface of thigh.

Terms used—thelarche—breast development, adrenarche—pubic hair, axillary hair growth, menarche—first menstrual period. Values in parentheses are mean age of appearance.

*% appearance and mean ages are for Indian affluent children (Indian Pediatrics 1992, Agarwal et al.)

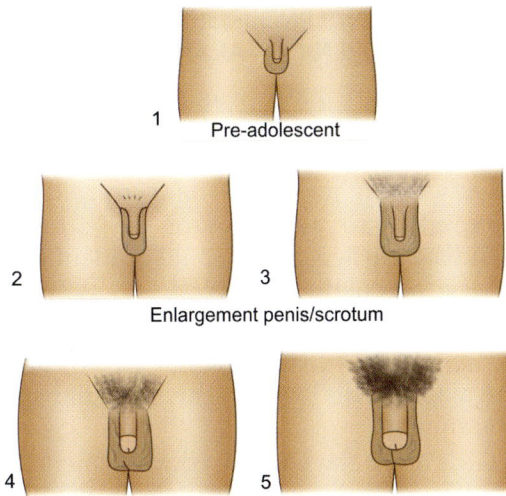

Fig. 2.2: *Stages of genital development (G 1–5) in boys [Tanner 1962]*

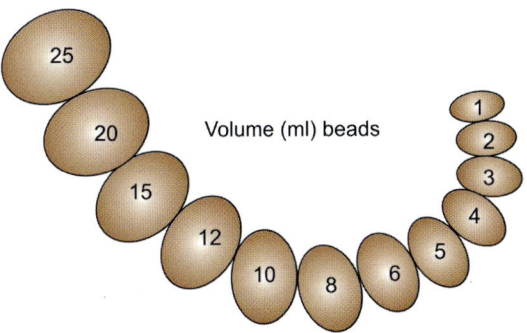

Fig. 2.3: *Orchidometer to measure testicular volume by Prader [Marshall and Tanner, 1985]*

Late maturer children finally may be taller than their peers as they will gain 5–6 cm/yr extra in linear growth in pre-pubescent growth.

Testicular volume is measured by Prader's orchidometer.

ACHIEVEMENT IN PHYSICAL GROWTH DURING PUBERTY

1. **Average duration of growth spurt:** 2.5–3 years.
2. **Total average gain in height**
   Boys : 27–29 cm
   Girls : 24–26 cm

3. **Height**
   At 10.5 years—height similar for boys and girls
   11.0–12.5 years—girls taller by 2 cm
   14.0 years—boys taller by 5 cm
   16.0 years—boys taller by 12 cm
4. **Height velocity**
   Girls: 9 cm/year (around B_2–B_3 stage) Menarche occurs after attainment of peak height velocity (PHV).
   Boys: 10.5 cm/year (around G_3–G_4 stage)
   During the year in which a boy attains his PHV he usually gains between 7 and 12 cm in stature, while a girl in the corresponding year gains 6–11 cm.
5. Total weight gain during adolescence is around 1/3rd of the adult weight (around 25–30 kg).
6. Peak weight velocity (PWV) (British children 9.8 kg/year) is preceded by PHV. Early maturing girls are more fatty and muscular as compared to the late maturers who are more lean.
7. **Changes in body proportions (under sex hormone effects)**
   a. Growth, begins in distal parts like feet and hands which also stop growing first. This is followed by the growth of arms and legs and finally trunk and chest. The growth of trunk stabilizes the further fall in US/LS ratio that normally occurs during childhood growth. Thus, the US/LS ratio is 1.1 at 10–11 years falls to 0.98–1.0 at 13–14 years, but finally stabilizes at 1.0–1.1 at completion of puberty.
   b. **Ratio of biacromial to bi-iliac dimensions**
   In boys is constant 1.37
   In girls decreases to 1.27
   c. Pelvic inlet is wider in girls because of growth of acetabulae. Male has greater stature, broader shoulders as compared to female who have wider hips.

8. **Head circumference:** In Indian affluent children, an increase of 1.5–2.0 cm in head circumference is observed during adolescent growth.
9. **Mid-arm circumference** also increases by 3.5 cm during entire adolescence.
10. **Face:** The greatest change in face is in the mandible. There is 25% of the total growth in height of the ramus between 12 and 20 years of age. There is loss of deciduous teeth and eruption of permanent teeth—cuspids, premolar and finally molars.
11. **Eyes:** Growth in the axial diameters of the eye results in an increasing tendency to "Myopia" in adolescence.
12. **Sebaceous glands:** Activity of sebaceous glands is responsible for acne development.
13. **Growth of larynx, pharynx and lungs** lead to the typical voice changes during adolescence.

Key Box

- Both primary and secondary sexual characteristics appear between 9 and 14 years of age in girls and 11 and 15 years of age in boys.
- Girls: Breast enlargement, occasionally initially unilateral, is the first obvious sign of puberty and occurs between 10 and 11 years of age.
- Boys: Penile and scrotal enlargement occur typically about a year after testicular volume increase—from 2.0 ml to >4 ml or testes length from 2 cm to 3.2 cm.
- Adolescence growth: Period extends from 2.5 to 3 yr; to cross SMR stages 2–5 (American kids take 4–5 yr).
- Height gain is 27–29 cm in boys and 24–26 cm in girls; weight gain in both 25–30 kg.
- Most rapid period of growth in the postnatal life (average gain being 19 g/day in boys and 16 g/day in girls).
- During this period skeletal growth is completed (50% of adult bone mass and 20% of the body stature). Mainly 'cortical bone growth' 1 cm height gain needs 20 g Ca. In adolescence 145 g Ca/yr is assimilated.

PUBERTY IN UNDERNOURISHED INDIAN RURAL CHILDREN OF VARANASI

- The 'Height gain' was similar to that attained by the affluent Indian children in adolescent growth spurt. Deficit of early life in height was not corrected.
- Weight gain was 38% of the affluent Indian. No age period could be identified for peak height velocity.

Boys had delayed maturation of
Genitals by 1.54 year; pubic hair by 0.82 year and axillary hair by 0.65 year, testicular vol. was similar.

- Girls had delayed breast development by 2.19 year and menarche was delayed by 0.82 year.

CAUSES OF DELAY IN PUBERTY ONSET; OR PUBERTY WITH NO PHYSICAL OR HORMONAL SIGNS

- Puberty may be delayed for several years and still occur normally—constitutional delay, a variation of healthy physical development (Fig. 2.5)
- Delay of puberty may also occur due to under nutrition as discussed above, many forms of diseases, i.e. tuberculosis, or to defects of the hypogonadism or the body's responsiveness to sex hormones.
- *In girls any of the following pointers:*
  i. No breast development by 13 years,
  ii. No menarche by 3 years after breast development
  iii. Menarche not attained by 16 years.
- *In boys any of the following pointers:*
  i. No testicular enlargement by age 14 years.
  ii. Pubic hair absent by age 15 years.
  iii. >5 years between the start and completion of growth of the genitalia

HEIGHT MEASUREMENT IN RELATION TO SEXUAL DEVELOPMENT

This understanding of sexual maturity to physical growth is important for pediatricians and parents, anthropologist, and

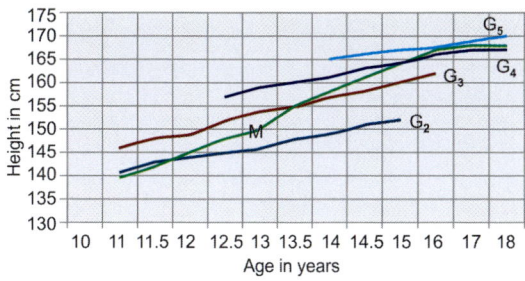

Fig. 2.4: *Height in relation to genital development and age in affluent Indian boys (M: Mean; G₂–G₅: Genital stages)*

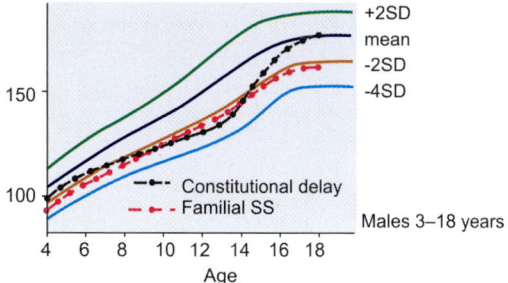

Fig. 2.5: *Catch-up growth, onset of puberty—constitutional delay*

medical practitioners as they use the distant growth curves in relation to age (Fig. 2.4, *see* example of 14 years boy showing median height as 158 cm and in stages SMR 2–5 height values are 151, 156.5, 162 and 166 cm, respectively). The height and weight measurements in relation to sexual development are given in Tables 2.2 and 2.3 (*see* on next page).

CHILDREN WITH CONSTITUTIONAL GROWTH DELAY (CGD)

The most common cause of short stature and pubertal delay, typically have retarded linear growth within the first 3 years of life. In this variant of normal growth, linear growth velocity and weight gain slows in the beginning as young as age 3–6 months, resulting in downward crossing of growth percentiles, which often continues until age 2–3 years. At that time, growth resumes at a normal rate, and these children grow either along the lower growth percentiles or beneath the curve but parallel to it for the remainder of the prepubertal years.

Comparison of the growth patterns between familial short stature and constitutional growth delay is shown in Fig. 2.5.

At the expected time of puberty, the height of children with constitutional growth delay begins to drift further from the growth curve because of delay in the onset of the pubertal growth spurt. Catch-up growth, onset of puberty, and pubertal growth spurt occur later than average, resulting in normal adult stature and sexual development (Fig. 2.5).

BRAIN DEVELOPMENT IN ADOLESCENCE

Teens' brains are not completely developed until late in adolescence. Studies suggest that the connections between neurons affecting emotional, physical, and mental abilities are incomplete. This could explain why some teens seem to be inconsistent in controlling their emotions, impulses, and judgments. Teens demonstrate a heightened level of self-consciousness. They tend to believe that everyone is as concerned with their thoughts and behaviors as they are.

There are four recognized psychosocial issues that teens normally deal with during their adolescent years. These include:

1. Establishing an identity
2. Establishing autonomy
3. Establishing intimacy
4. Becoming comfortable with one's sexuality.

Key Developmental Experiences

The process of adolescence is a period of preparation for adulthood during which time several key developmental experiences occur. Besides physical and sexual maturation, these experiences include movement toward social and economic independence, and development of identity, the acquisition of skills needed to carry out adult relationships and roles, and the capacity for abstract reasoning. While adolescence is a time of tremendous growth and potential, it is also a time of considerable risk during which social contexts exert powerful influences.

Table 2.2: *Weight and height percentiles in relation to genital development and age in boys*

Age yrs		3	10	25	50	75	90	97	Numbers
					Genital development stage (SMR-2)				
10	Wt.	23.0	25.3	27.5	31.1	35.0	40.8	48.2	259
	Ht.	128.2	130.8	135.5	139.7	144.6	148.5	151.3	
11	Wt.	24.5	134.7	29.4	32.6	37.0	42.9	51.0	480
	Ht.	130.7	13.7	138.5	142.4	146.8	151.0	155.3	
12	Wt.	26.1	28.1	31.0	34.9	39.9	46.3	55.8	518
	Ht.	133.8	137.3	141.7	145.5	150.0	154.5	159.3	
13	Wt.	28.2	30.2	32.5	36.5	41.5	52.2	61.2	287
	Ht.	136.5	141.0	145.0	149.0	152.2	156.0	160.6	
14	Wt.	31.4	32.2	35.0	38.9	42.5	49.0	62.1	91
	Ht.	142.4	145.7	148.6	153.4	156.6	159.9	166.6	
					Genital development stage (SMR-3)				
10	Wt.	23.8	26.2	28.9	34.7	40.2	45.4	49.1	64
	Ht.	127.8	134.6	137.8	145.6	151.8	155.1	159.1	
11	Wt.	28.4	29.7	32.8	36.4	41.8	46.7	50.9	165
	Ht.	134.2	139.6	143.6	148.0	153.1	158.5	163.0	
12	Wt.	28.7	32.4	35.4	39.0	44.2	50.6	59.4	366
	Ht.	139.6	143.4	148.2	152.2	156.4	160.5	163.0	
13	Wt.	30.4	32.7	35.8	39.9	45.6	51.7	61.9	467
	Ht.	141.5	142.2	149.5	154.1	158.5	162.1	166.9	
14	Wt.	32.8	34.7	38.2	42.8	48.6	54.4	64.8	311
	Ht.	145.7	149.3	153.5	157.8	161.8	165.6	169.1	
15	Wt.	37.2	38.0	42.0	46.1	51.3	61.6	77.0	116
	Ht.	152.5	152.5	156.1	161.1	166.0	169.3	172.7	
16	Wt.	39.6	41.8	44.1	50.2	55.3	63.7	71.9	45
	Ht.	154.6	157.0	157.4	162.0	165.6	172.8	179.6	
					Genital development stage (SMR-4)				
11	Wt.	27.9	35.8	39.7	43.2	49.0	54.3	58.6	28
	Ht.	138.8	147.4	150.9	156.4	161.3	163.0	166.0	
12	Wt.	32.1	35.1	38.4	430.	47.7	54.1	61.0	156
	Ht.	143.0	148.4	153.1	157.9	162.5	167.0	168.9	
13	Wt.	33.8	36.8	41.1	45.6	51.1	59.4	71.1	430
	Ht.	147.0	151.5	156.0	160.2	165.0	168.5	172.6	
14	Wt.	34.8	38.5	42.5	47.7	52.7	58.9	68.6	555
	Ht.	148.5	154.0	158.0	162.6	167.5	171.2	174.8	
15	Wt.	39.2	41.7	45.2	50.8	56.5	65.7	75.2	407
	Ht.	154.0	157.0	161.0	165.5	169.3	172.8	176.3	
16	Wt.	41.8	43.5	47.8	52.6	60.2	69.5	76.6	210
	Ht.	156.7	159.0	162.5	166.5	171.5	175.0	178.9	
17	Wt.	43.9	45.7	48.4	55.1	62.9	68.6	76.2	68
	Ht.	160.3	161.0	164.1	167.8	173.7	177.5	184.3	
					Genital development stage (SMR-5)				
13	Wt.	39.3	41.3	45.8	50.5	57.2	65.7	71.3	80
	Ht.	152.9	157.6	160.5	164.9	169.1	173.0	177.4	
14	Wt.	40.5	42.9	46.8	52.0	57.7	65.8	77.3	316
	Ht.	153.7	158.0	162.0	166.0	170.7	174.1	178.2	
15	Wt.	41.6	44.0	47.3	52.3	59.6	69.2	77.9	461
	Ht.	156.5	159.2	163.5	167.0	171.5	174.5	179.4	
16	Wt.	42.5	45.4	50.0	55.0	62.5	70.5	82.3	420
	Ht.	157.9	160.2	164.7	168.9	173.5	177.2	181.5	
17	Wt.	44.8	47.1	52.3	57.1	63.1	70.9	80.6	259
	Ht.	160.3	162.0	165.4	169.3	173.3	176.5	181.0	
18	Wt.	43.9	46.5	51.7	58.3	64.5	74.7	84.9	75
	Ht.	157.5	163.8	166.5	170.0	173.0	177.9	183.2	

Indian Pediatr 2001;38:1217–37 (Agarwal et al). Wt: weight; Ht: height

Table 2.3: *Weight and height percentiles in relation to genital development and age in girls*

Age yrs		Percentiles							Numbers
		3	10	25	50	75	90	97	
Breast development stage (SMR-2)									
10	Wt.	23.8	26.4	29.3	32.3	36.5	42.3	46.9	383
	Ht.	127.7	131.5	135.4	139.4	143.9	147.3	151.0	
11	Wt.	24.5	26.7	29.0	32.2	36.8	42.1	50.6	353
	Ht.	131.1	134.3	137.7	142.5	146.2	149.7	154.7	
12	Wt.	25.8	27.2	29.9	33.2	37.4	43.0	48.7	195
	Ht.	134.9	137.6	140.0	144.5	149.1	152.6	157.7	
13	Wt.	27.0	29.1	30.7	36.1	41.0	46.8	50.9	53
	Ht.	139.9	141.3	144.9	149.5	153.7	161.0	165.2	
Breast development stage (SMR-3)									
10	Wt.	29.0	31.2	33.1	37.6	43.0	48.9	53.0	131
	Ht.	134.9	138.5	141.0	144.3	147.5	150.5	153.7	
11	Wt.	30.1	32.4	34.8	38.8	43.5	48.3	56.2	316
	Ht.	138.0	140.0	143.2	147.0	150.5	154.5	157.1	
12	Wt.	30.9	33.6	35.9	39.4	44.5	51.2	57.2	330
	Ht.	140.0	143.5	146.0	150.2	153.5	157.3	162.0	
13	Wt.	32.0	33.5	36.3	40.0	45.0	50.6	60.6	242
	Ht.	141.5	145.0	148.0	151.9	156.7	160.1	163.3	
14	Wt.	32.4	34.2	38.8	42.9	47.0	52.1	58.3	143
	Ht.	144.0	145.8	150.0	154.2	157.0	161.7	164.5	
15	Wt.	34.8	37.8	39.1	43.0	46.5	48.2	53.2	71
	Ht.	146.5	149.0	150.6	155.0	157.3	162.7	164.9	
16	Wt.	36.2	37.2	38.2	43.0	46.8	52.6	55.2	34
	Ht.	147.1	148.9	152.0	155.2	158.4	163.5	168.3	
Breast development stage (SMR-4)									
11	Wt.	29.9	35.5	39.3	43.6	49.6	54.9	58.7	115
	Ht.	138.2	142.7	147.0	150.3	154.5	156.3	161.7	
12	Wt.	34.4	37.8	40.2	44.5	49.3	53.9	59.1	244
	Ht.	142.7	145.3	148.5	152.0	156.0	159.3	161.9	
13	Wt.	33.4	37.6	40.5	44.3	49.6	55.3	64.3	304
	Ht.	144.0	146.2	150.1	153.3	157.1	160.6	164.2	
14	Wt.	35.7	38.0	40.9	44.4	50.3	55.3	63.6	264
	Ht.	145.1	146.5	150.5	154.5	157.7	161.6	167.0	
15	Wt.	35.7	38.7	42.3	45.8	50.4	56.0	64.5	142
	Ht.	146.2	148.2	152.0	155.0	159.0	161.7	167.7	
16	Wt.	38.2	39.5	43.5	48.0	53.3	57.3	66.4	89
	Ht.	147.1	148.7	152.7	156.0	159.9	163.0	167.1	
17	Wt.	37.4	39.7	43.0	47.7	51.5	54.2	59.3	44
	Ht.	148.0	149.3	152.0	155.5	159.0	165.0	166.8	
Breast development stage (SMR-5)									
12	Wt.	39.0	42.7	47.3	52.8	58.6	63.3	66.1	88
	Ht.	142.8	147.5	149.8	152.6	156.8	161.1	163.2	
13	Wt.	39.4	42.0	45.7	51.8	59.3	66.1	71.5	178
	Ht.	143.0	147.5	150.5	155.1	160.5	163.8	167.8	
14	Wt.	37.1	41.5	45.4	51.4	57.3	65.0	73.2	189
	Ht.	145.6	147.8	151.5	155.0	159.5	162.2	165.6	
15	Wt.	36.5	41.5	44.9	49.9	58.5	66.2	72.4	187
	Ht.	146.2	148.1	151.5	156.0	160.8	164.5	168.0	
16	Wt.	38.4	41.8	46.4	50.5	58.0	64.8	75.2	148
	Ht.	147.1	148.2	152.8	156.6	159.9	163.0	166.9	
17	Wt.	41.2	43.4	48.0	53.2	57.7	62.8	77.0	60
	Ht.	147.5	148.8	153.2	157.2	160.4	163.8	168.1	

Indian Pediatr 2001;38:1217–35 (Agarwal et al).

Adolescents are not fully capable of understanding complex concepts, or the relationship between behavior and consequences, or the degree of control they have or can have over health decisions including those related to sexual behavior.

The Adolescent Brain

Time-lapse MRI images of human-brain development between ages 5 and 20 years.

- Show the gray matter growth and then gradual loss. Paradoxically, the thinning of gray matter that starts around puberty corresponds to increasing cognitive abilities. Improved neural organization and increase in the white matter, which helps brain cells communicate.

Brain growth in adolescence. National Institute of Mental Health. *E-mail: nimhinfo @nih.gov*

Secret of the Teen brain: Inside the adolescent brain *http://www.time.com/time/ health/article/0,8599,1919663,00. html#ixzz25hOo8Gfz* and see Chapter 9 on "Plasticity in Developing Brain".

Box 2.2: *Adolescent brain*

- Brain adopts a *"use-it-or-lose-it" pruning system*, sloughing unused connections and increasing the speed of others. *30,000 synapses may be lost per second in the early adolescent brain leading to an ultimate loss of almost one half of the synapses*
- Areas of the brain responsible for *executive functioning* (such as *strategic thinking, weighing risks and benefits and impulse control*) continue to develop and refine connections through adolescence and into the mid-twenties

---

Box 2.1: *Children's brains are much busier than an adult's*

- **Gray matter** is made up of the cell bodies of neurons, the nerve fibers that project from them, and support cells
- At **birth** each neuron has 2,500 synapses
- By **2 years**, there are 15,000 synapses per neuron
- At **3 years** the first period of consolidation begins. This period tends to be characterized by children asking the question "why?"
- It is estimated that a **four-year**-old asks a "why" question every two and a half minutes!
- Around the age of **six**, there is a second surge as the brain starts to use language in increasingly complex ways
- Up to the age of **nine** a child's brain continues to **be twice as active as an adult's brain.**

Box 2.3: *Alcohol and the adolescent brain*

- Delays in normal brain development over time
- More vulnerable to long-term damage to memory and other systems
- Prevents changes in neural circuits involved in learning and attention
- Prone to seizures after binge drinking
- Less brain activity overall
- Less vulnerable to perceived negative effects: Motor coordination and sedation
- Reduced hippocampus volumes in alcohol abuse indicates permanent brain damage
- Causes reduced testosterone in adulthood

Tobacco and brain

Smoking saturates receptors: As nicotine from a cigarette attaches to the 04β2* -nACH nicotinic receptors in the brain, it displaces a radiolabeled tracer (red and yellow indicate high levels of the tracer, green indicates intermediate levels, and blue indicates low levels). The nicotine from three puffs displaced 75 percent of the tracer from study participants' receptors, and the nicotine from three cigarette, nearly all.

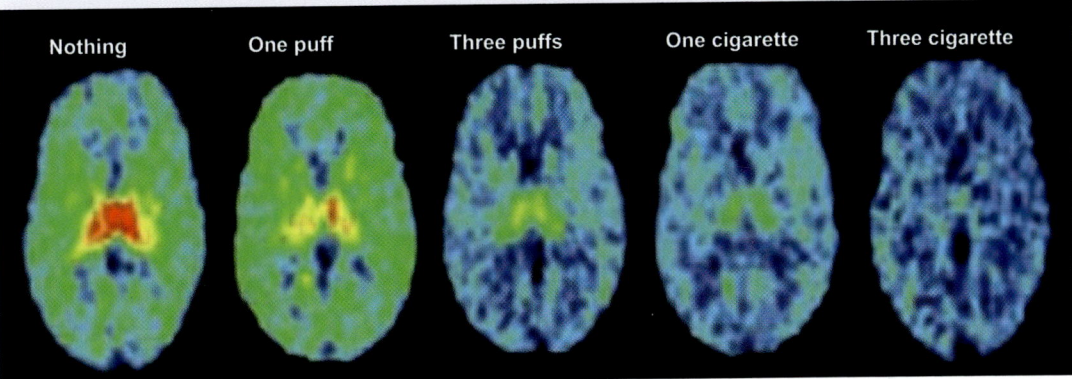

Nothing   One puff   Three puffs   One cigarette   Three cigarette

Fig. 2.6

Reference: For further reading on brain in adolescence (TIME-10th May 2004. Secrets of the Teen Brain).

HEALTH SCREENING DURING ADOLESCENCE

1. Measure height, weight and calculate BMI (body mass index).

2. Assess sexual development—Give photographs for Tanners sexual development stages to the child and let her/him mark the stage. Ask him to feel for testes if both are palpable mark. In school surveys we had detected undescended testis/hernia/hydrocele/hypospadias, etc.

3. Please check the measurements against Sexual Maturity Grade in the tables given above (Tables 2.1 and 2.2).

4. Measure blood pressure, vision in both eyes and hearing.

5. Advise dental health screening and oral hygiene.

6. General examination to see that there is no abnormality—check for anemia any bony abnormality.

7. Diet and exercise

EXERCISE FOR 40 MINUTES DAILY

Why nutrition for 5–10 years growth is very important?

- Middle childhood, is characterized by a slow, steady rate of physical growth. However, cognitive, emotional and social development occur at a tremendous rate.
- To achieve optimal growth and development, children need a variety of healthy foods that provide sufficient energy, protein, carbohydrates, fat, vitamins and minerals. They need three meals per day, plus snacks (working mother with school kids).
- Preparatory period for adolescence. Growth failure of this period (undernutrition) will delay onset of pubscence; No catch-up growth; further the PHV will not be observed.
- Approximate growth—Ht 25 cm girls and 30 cm in boys; wt 15–30 kg.
- Lymphoid growth is maximum in this period (immunity).

Table 2.4: *Dietary recommended intake 30th Nov 2010 (IOM, USA)*

Age yr	Protein g/day	CHO g/day	Calcium mg
4–8 Boys 1200 to 2000 cal	19	130	1000 mg (Vit D = 600 IU)
Girls 1200 to 1800 cal			
Boys 9–11 1600 to 2600 cal	34	130	1300 mg (Vit D = 600 IU)
Girls 9–11 1400 to 2200 cal			
Boys 14–18 2000 to 3200 cal	52	130	Same—1300/600
Girls 14–18 1800–2400 cal	46	130	Boys have more lean body mass
Sexual development	60	3000 kcal/d	2 times of girls—need more
Boys (<12 years)	46.0	2500 kcal/d	Fe, Ca and Zn

SUGGESTED READING

1. Agarwal DK, et al. Physical and sexual growth pattern of affluent Indian children from 5 to 18 years of age. Indian Pediatr 1992;29:1203–82.

2. Agrawal KN, et al. Physical growth assessment in Adolescence. IBID 2001;38:1217–35.

3. IAP Textbook of Pediatries, 2013; 5th ed. p. 90–96.

4. Nelson Textbook of Pediatries, 2011 19th ed. p. 649–63.

3

Common Stature Variations

Vijayalakshmi Bhatia

SHORT STATURE

Optimal growth of children is of prime concern to the parents. A common reason for seeking pediatric consultation is a perceived shortness of the child. In this situation, a pediatrician has to decide whether the child is actually short, and if yes, whether the short stature is a variation of the normal (thus not needing extensive investigative work up) or is it pathological. More importantly, a pediatrician should be able to pick up children with growth faltering even before they present with obvious shortness. Knowledge of normal growth during childhood and adolescence is essential for carrying out this evaluation (*see* Chapters 1 and 2).

This section outlines the approach to a short child. The subject is dealt with under the following sub-headings:

1. Definition of short stature
2. Selection of children requiring further evaluation
3. Etiology of short stature
4. Evaluation of a short child
5. Management approach

Who is a Short Child?

The definition of short stature is arbitrary. The general rule is that any child whose height falls below the 3rd centile for his/her age and sex is considered to be short. Thus, 3% of children would be classified as being short whether or not there is anything wrong with them. However, about 10–15% among these would have a pathological cause for their shortness.

In addition, a child is also considered to be short if his height is below his genetic potential, even if his height is above the age and sex specific 3rd centile. The genetic potential for growth is determined by the mid-parental height (MPH), the average of parental height. Since men are genetically pre-determined to be taller than women by about 13 cm, one must add 13 cm to mother's height before averaging it with father's height to calculate the mid-parental height for a boy. Likewise, while calculating mid-parental height (MPH) for a girl, 13 cm need to be subtracted from the father's height. Thus, the formula for calculating mid-parental height is as follows:

MPH for boys

$$= \frac{\text{Father's height} + \text{mother's height} + 13 \text{ cm}}{2}$$

MPH for girls

$$= \frac{\text{Father's height} - 13 \text{ cm} + \text{mother's height}}{2}$$

The mid-parental height is plotted on child's growth chart, on the extreme right column line, corresponding to 18 years. Next, the child's height is plotted on his/her age column. The projected height is determined by extrapolating his growth along his growth channel to anticipated adult height. If a child's projected height is within 5 cm of the mid-parental height (target height), the height is appropriate for his genetic potential. However, if the projected height and target height differ by more than 5 cm, the child needs to be evaluated for a pathological cause for shortness. However, it should be kept in mind that many diseases like skeletal dysplasias and some endocrine and metabolic disorders characterized by short stature may also be genetically transmitted. Thus, mere identification of short stature as genetically transmitted does not always exclude an underlying pathological cause.

Yet another group of children who need evaluation are those whose growth velocity is sub-normal. These children may not be short as yet, but may become so if remedial measures are not taken at this stage. Growth velocity is determined by sequentially measuring and plotting height at 3 monthly intervals for at least 6–12 months (minimum three measurement points; calculate velocity in cm/yr). Since growth usually occurs in spurts, a shorter period of observation will not suffice.

Normal growth velocities during childhood are presented in Table 3.1. A broad principle to be kept in mind is that between 2 years and puberty, children grow with remarkable fidelity relative to their chosen growth channel. Any crossing of height centiles during this age is significant and should be evaluated.

Children who Require Evaluation

From the above discussion it is gathered that the following groups of children need further evaluation:

Table 3.1: *Normal growth velocity during childhood*

Birth to 1 year	: 25 cm/year
1 to 2 years	: 12 cm/year
2 to 3 and 3 to 4 years	: 6 to 7 cm/year
4 years till pubertal onset	: 5 cm/year

a. Those with height below the age and sex specific 3rd centile.
b. Those with height below the genetic potential.
c. Those with a poor height velocity (Table 3.1)

Etiology of Short Stature

Causes of short stature in children are presented in Fig. 3.1. Broadly, shortness may be normal, or pathologic. The former includes children who are short, but healthy and growing normally, whose shortness is of genetic origin. This group must be differentiated from those with a pathologic cause for shortness/poor growth so that appropriate remedial action can be taken where possible.

Normal Variants

a. Familial short stature

Children with this condition have short parents and the child's height is commensurate with the genetic potential. Children with familial short stature are either born small, or cross their growth channel downwards during 6–18 months of age. Subsequently, they assume a place somewhere below (but close to) the 3rd centile and continue to grow along that channel with a normal growth velocity. Their bone age is not retarded and puberty occurs at the normal time. The final stature is small but consistent with the family.

b. Constitutional delay in growth and puberty (CDGP); see Page 37, Fig. 2.5

Children with this condition are born normal in size and grow normally for the first few months. However, the growth

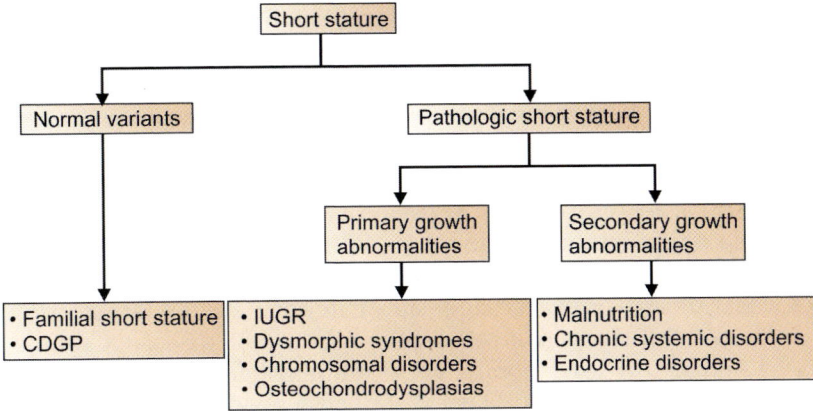

Fig. 3.1: *Etiology of short stature (CDGP constitutional delay of growth and puberty)*

slows for no apparent reason between 6 and 36 months when both height and weight gain falter. There is no systemic illness and the nutritional status of these children is otherwise normal. Subsequently, the growth velocity normalizes and these children continue to grow parallel to, but below the normal growth curves. The bone age is delayed compared to chronological age but equals height age. This is also reflected in a delayed onset of puberty, but the final sexual development is normal. The adult height is in the low normal range.

CDGP has a strong familial incidence. History of delayed periods in mother or delayed height spurt in father may be a clue towards the diagnosis. The etiology of CDGP is not clear.

Primary Growth Abnormalities

Included in this category are disorders of growth where the defect(s) appears to be intrinsic to the growth plate:

a. Intrauterine growth retardation (IUGR)

Though most IUGR babies exhibit catch-up growth during infancy, about 10% do not. Their growth velocity during childhood is usually normal. These children may also have an early puberty and are unusually short as adults. Included in this category are some children with severe IUGR and others who along with short stature also have dysmorphic features. The latter group includes disorders like Russell-Silver syndrome, Seckel syndrome, etc.

b. Chromosomal abnormalities

Certain chromosomal defects, notably Down and Turner syndromes are also associated with short stature. While the former is usually clinically obvious, the latter may not always be so, especially when it results from chromosomal mosaicism. Thus, every girl with unexplained shortness must have a karyotyping to exclude Turner syndrome. The children with Down and Turner syndromes also have a low height velocity. The height deficit is further exaggerated during adolescence when pubertal height spurt is delayed and often poor.

c. Osteochondrodysplasias

These disorders are uncommon, but are an important cause of extreme short stature, usually of a disproportionate type. Family history may be positive, though some cases may occur as fresh mutations. Abnormalities of limbs and trunk which may be obvious on clinical examination are confirmed by careful anthropometry including measurement of upper/lower segments (US/LS), sitting height, arm span and head circumference. As an initial screening, abnormality of US/LS ratio provides a useful pointer towards presence of these disorders. (Normal US/LS ratios are: birth = 1.7; 6 months = 1.6; 1 year = 1.5; 3 years = 1.3 and 7 years – 1.0). A classic example of a disorder with high US/LS ratio (short limbed dwarfism) is achondroplasia, while low US/LS ratio (short trunk dwarfism) is found in spondyloepiphyseal dysplasia and mucopolysaccharidosis. The exact diagnosis of the type of skeletal dysplasia requires radiologic evaluation of long bones, vertebral column, pelvis and the skull.

Secondary Growth Disorders

a. Malnutrition

Globally, the most common cause of growth stunting is malnutrition. In India, as per the National Family Health Survey-III (2005–2006), 20% of children <5 years of age had wasting (weight-for-height below –2SD by WHO Child Growth Standards) while 45% had stunting (height-for-age < –2SD by WHO Child Growth Standards) *see* Chapter 12. It is prudent to reiterate here that growth during the first 6 months of life is independent of GH and is largely dependent on the nutrition. Poor nutritional intake during this period of rapid growth can have profound and long lasting effect. Once GH dependent growth is established, normal growth ensues, but stature lost at this stage is not easily recovered. Likewise, continued chronic malnutrition during childhood leads to stunting which is not always fully recoverable. An important clinical pointer towards malnutrition being the underlying cause of short stature is that the weight is at least as severely, and usually more severely affected than the height.

b. Chronic systemic disorders

Chronic illness involving a variety of systems is another common and important cause of growth failure. The children when seen may or may not be short, but their height velocity is slow. Some of these ill-nesses have other major manifestations and may already be known to the family. The examples include congenital heart disease, specially the cyanotic ones, chronic anemia specially thalassemia and sickle cell anemia, chronic asthma, cystic fibrosis, chronic renal insufficiency (CRF), etc. However, other illnesses may not be obvious and present primarily as short stature. Important in this category are malabsorption states, especially celiac disease, Crohn's disease, renal tubular acidosis (RTA) and sometimes, CRF. These need to be specially investigated for in children with unexplained short stature.

c. Endocrine disorders

The most important endocrine disorders presenting as short stature include hypothyroidism and GH (growth hormone) deficiency. Long standing glucocorticoid excess, whether iatrogenic or secondary to an ACTH/corticosteroid producing tumor, uncontrolled diabetes, severe rickets and pseudohypoparathyroidism are other endocrine causes of poor growth.

Congenital hypothyroidism is easily recognized by the classic findings of short stature, delayed development, constipation and coarse facial features (Fig. 3.2). Acquired hypothyroidism is less obvious, and presents primarily as poor physical growth, weight gain, lethargy and constipation. A goiter may often though not invariably be present.

GH deficiency (GHD) may be isolated or occur as a part of panhypopituitarism. It may be congenital which is more likely, or acquired, secondary to pituitary destruction due to tumor, trauma, infection, etc. It may also be due to hypothalamic deficiency due to, for example, a craniopharyngioma or hydrocephalus. The characteristic picture of severe GH deficiency is a very short child with a round immature face (Fig. 3.3). Craniofacial midline deformities, if present, are a useful clinical pointer. Many babies with GHD have normal birth weight and length, poor growth rate being apparent only in infancy or childhood. Small penis in a male child is a clue towards GHD.

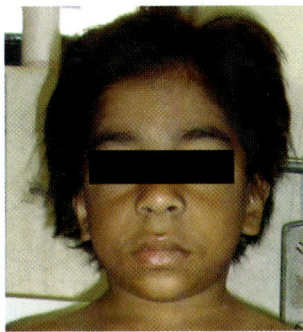

Fig. 3.2: *Clinical features of hypothyroidism, in the form of thickening of the lips and nose, puffiness around the eyes and downy hair on the forehead, can be appreciated in this girl who presented for the complaint of short stature*

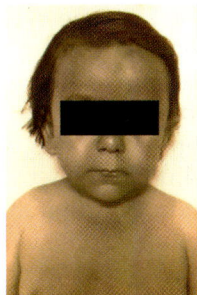

Fig. 3.3: *An 8-year-old child with growth hormone deficiency exhibits frontal bossing, a depressed bridge of the nose, midfacial hypoplasia and a doll-like appearance*

Evaluation of a Child Presenting with Short Stature

Key points to be considered in the clinical assessment of a child with short stature are presented in Table 3.2. One should begin with asking about birth weight and length, to look for intrauterine growth retardation (IUGR). Delayed development, a cardinal feature of congenital hypothyroidism, is absent in patients with isolated GH deficiency and skeletal dysplasias, other important causes of profound shortness. It is important to assess whether the retardation is in height alone, or in both height and weight, since most systemic disorders and specially malnutrition affect weight more than the height. Endocrine conditions, on the other hand, have a much more profound effect upon the height.

In adolescents, stature is evaluated in relation to the pubertal state, since sexual maturity is associated with a height spurt (*see* Chapter 2). Thus, delayed puberty with associated delay in height spurt may cause apparent shortness in a child. On the other hand, a child with precocious puberty has an early height spurt, but may appear shorter when her peers are experiencing a height spurt and will be short as an adult. Puberty is delayed in CDGP, systemic illnesses and endocrine disorders, while it is normal in children with familial shortness, skeletal dysplasias and children with IUGR.

Table 3.2: *Clinical evaluation of a child with short stature*
History
• Birth weight, length, gestation
• Retardation in height alone, or both height and weight
• Dietary history
• Developmental history
• Systemic review
• Past treatment/hospitalization
• Family history—short stature, delayed puberty, endocrine illness, syndromes
• A review of previous height and weight measurements, if available

Contd.

Table 3.2: *Clinical evaluation of a child with short stature (Contd.)*

Examination

- Anthropometry: Height/supine length, weight, US/LS ratio, arm span-height difference.
- Facies, midline defects, micropenis, thyroid enlargement, rickets, truncal obesity, dysmorphic features.
- Assessment of nutritional status
- Systemic examination
- Pubertal assessment
- Height of parents and siblings

Management

Investigations

Table 3.3 summarizes the investigative approach to a short child.

As accurately assessed bone age is an important tool in the evaluation of a child with short stature. It not only provides a clue regarding the etiology of shortness, but also gives information about the remaining growth potential. The bone age is delayed but is commensurate with the height age in a child with CDGP and various systemic conditions leading to short stature. A profound delay in bone age, even more than the retardation in height age is found in hypothyroidism and GH deficiency. Bone age is typically normal in children with familial short stature (*see* Chapter 6).

A child with skeletal abnormalities on clinical examination and radiography should be evaluated for metabolic bone diseases like mucopolysaccharidosis, mucolipidosis, etc. Every child with no obvious cause for shortness should be investigated to exclude celiac disease, renal tubular acidosis, chronic renal disease and in case of girls, Turner syndrome. GH assay should be carried out in a child with clinical suspicion of GH deficiency once other possible causes have been excluded and thyroid function tests (including T4, as TSH will not be diagnostic in central hypothyroidism)have been found to be normal. Many Western textbooks suggest estimation of insulin like growth factor-1 (IGF-1) and IGF binding protein-3 (IGFBP-3) as screening tests for GH deficiency. However, these estimations are difficult to interpret due to wide variation of their normal levels and overlap between normal and GH deficient children. Moreover, these are

Table 3.3: *Investigations in a child with short stature*

Basic investigations

- Bone age estimation
- Hemogram
- Serum biochemistry—liver function, renal function, electrolytes, proteins, fasting blood sugar
- Venous blood gases
- Urine pH, specific gravity, proteins, casts
- Stool for giardiasis
- X-ray skull for sella and supra-sellar calcification

Second line investigations

- Karyotype in all girls with unexplained shortness
- Tissue transglutaminase antibody/jejunal biopsy to exclude celiac disease
- Tests for renal tubular acidosis
- Provocative GH testing

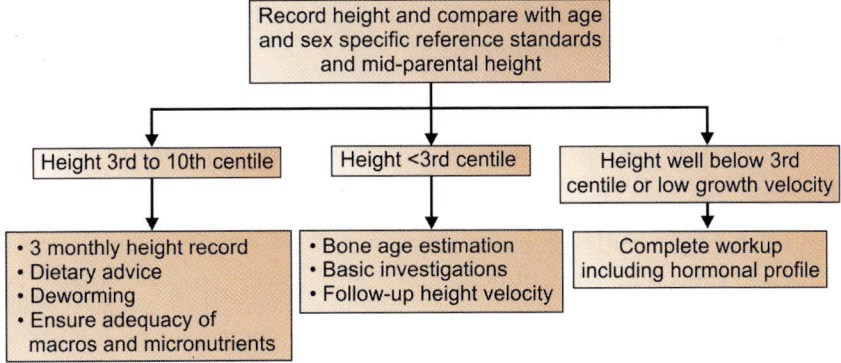

Fig. 3.4: *Management approach to a short child*

expensive. Thus, they cannot be recommended as a screening test in our set up.

In a child with suspected GH deficiency estimation of basal GH levels is of limited utility since GH is released in a pulsatile manner and spontaneous GH secretion is low during many hours of the day. Thus provocative tests designed to evaluate the integrity of hypothalamopituitary axis are carried out. Insulin-induced hypoglycemia stimulation test has been the gold standard of pharmacological assessment of GH secretion but is fraught with the risk of hypoglycemic CNS complications. The clonidine stimulation test, is therefore, a preferred test, wherein 150 ugm/sq m surface area is given early morning orally. Blood sample for GH estimation is collected at 0 min and 30, 60, and 90 min after clonidine. Peak stimulated serum GH level of <5 ng/ml, traditionally demonstrated with at least 2 pharmacological agents (unless there is a clinical setting of GH deficiency such as a very typical phenotype or a pituitary/hypothalamus tumor in which case 1 test should suffice) is suggestive of complete and 5 to 10 ng/ml of partial GH deficiency. Other agents used for GH provocative studies include L-dopa, glucagon and arginine. Finally, in a child with clinical features of GH deficiency but with normal/increased serum GH levels.

GH resistance syndromes should be ruled out. Figure 3.4 summarizes the management approach to a short child.

Treatment

Treatment depends upon the cause of short stature. Patients with familial short stature do not require any therapy. Those with primordial dwarfism and most skeletal dysplasias are unlikely to benefit from any intervention. However, the role of appropriate counseling in these conditions cannot be undermined. When growth failure is the result of a chronic systemic disease, correction of the primary medical problem would result in catch up growth and improvement of final stature. However, if glucocorticoids are being used for management of the underlying condition, growth failure may be profound and is unlikely to be corrected until the patient is weaned off the steroids.

Subjects with constitutional delay in growth and puberty (CDGP) also require counseling regarding the benign and transient nature of the disorder and reassurance about the final height status which however, is likely to be along the lower end of the normal height for family. No treatment is required in most cases. However, in subjects with tremendous anxiety, short-term low dose sex steroids may help by tiding over the problem till natural pubertal progression ensues. In males, this is given as a 4 weekly injection of testosterone enanthate 50 mg intramuscularly for 4–6 months. Girls may be given ethinyl estradiol 2.5 mcg daily for the same period. The hormone supplement should then be stopped and the patient re-evaluated to ascertain that 'true' puberty has ensued. If not, a repeat course can be considered. Failure to enter into spontaneous puberty beyond 1 year of hormone replacement warrants evaluation for pituitary/gonadal insufficiency.

Hypothyroid subjects show a good catch up growth, once replacement with thyroxine is started. However, with thyroxine, skeletal maturation also advances rapidly, thereby potentially limiting the final height in cases with long standing hypothyroidism and associated profound height deficit.

Patients with GH deficiency require GH replacement. The therapy is initiated at a low dose and titrated slowly upwards. The dose is 25–50 ug/kg/day as subcutaneous injections. The treatment is ideally continued until final height or epiphyseal closure has been documented. The therapy is however, expensive, the approximate cost being ₹ 1.5–2.0 lakhs/year for a 20 kg child. GH is best administered under the expert care of an experienced pediatric endocrinologist.

GH therapy also improves the height outcome in patients with chronic renal failure, Prader-Willi syndrome, Turner syndrome and IUGR children who show no natural catch-up growth by 3 years of age. However, the quantum of benefit is small and unpredictable compared to the results of GH treatment for children with GHD, and often higher doses are required than that needed for GHD therapy. Therefore, parents and treating pediatrician must have realistic expectations from such therapy.

Key Box

- A careful anthropometric assessment and repeated height measurements is the key tool in initial evaluation of a short child.
- The growth of any child should be evaluated with reference to age and sex specific standards as well as the parental height.
- Most cases of short stature are due to normal growth variants. Chronic systemic diseases are an important cause in the remaining.
- Accurate bone age evaluation is the basic investigation which not only indicates the underlying cause of shortness but also gives information regarding the remaining height potential.
- Important endocrine conditions leading to short stature include hypothyroidism, GH deficiency, rickets, Turner and Cushing syndromes.

TALL STATURE

Unlike shortness, tall stature is a relatively rare concern in pediatric practice. The most common cause is familial tall stature. There are some primary syndromes associated with tall stature. Secondary causes of tallness are rare. Table 3.4 lists the conditions that can lead to tallness/large size.

(i) Normal variants

a. **Familial tall stature:** A child with familial tallness is tall during the growth period and has an increased final height. The final height is within the range defined by parental size. They have a normal height velocity (at approximately 75th centile) and puberty begins at the normal time. The bone age is consistent with the chronological age.

b. **Tall stature with nutritional obesity:** Children with simple obesity are taller as compared to their peers and have a height at the upper end of their predicted genetic potential. However, their weight centile is higher than their height centile. These children also often have an early puberty.

Table 3.4: *Etiology of tall stature/large size*

I. Normal variants

- Familial tall stature
- Constitutional early puberty
- Tall stature with nutritional obesity

II. Primary causes of tall stature

- Proportionate large size with intellectual deficit: Sotos syndrome (cerebral gigantism), Beckwith-Wiedemann syndrome, Marshall-Smith syndrome, Weaver syndrome.
- Disproportionate tall stature with normal intellect: Marfan syndrome, hypogonadism
- Disproportionate tall stature with intellectual deficit : Klinefelter syndrome (XXY, XXYY, XXXY and mosaics), the XYY syndrome, homocystinuria

III. Secondary causes of large size

- Growth hormone excess.

(ii) Primary causes of tall stature

a. **Proportionate large size with intellectual deficit:** Several syndromes may be included in this category, the most important being **Sotos syndrome** (cerebral gigantism). Children with this disorder have dysmorphic features including prominent forehead, high arched palate, hypertelorism and pointed chin with associated developmental retardation and intellectual handicap. They show a rapid growth in infancy that persists during the first 3–4 years, but they are not tall as adults. Their bone age is advanced by 1–2 years. The diagnosis depends upon the presence of characteristic features. No specific management is indicated.

Beckwith-Wiedeman syndrome is generally recognized in the neonatal period on the basis of somatic overgrowth and hypoglycemia. The babies are big at birth and have macroglossia, craniofacial abnormalities and visceromegaly.

Other uncommon syndromes associated with tall stature include **Marshall-Smith** and **Weaver** syndromes.

b. **Disproportionate tall stature with normal intellect:** Patients with **Marfan syndrome**, a heritable disorder of connective tissue fall into this category. The clinical features include disproportionate tall stature with long arms and legs, long fingers and toes (arachnodactyly), joint laxity, hernias, scoliosis and chest abnormalities, ocular defects are frequent and include high myopia and dislocation of lens. Cardiac defects specially mitral and aortic regurgitation may be present in these patients and are an important cause of reduced lifespan.

Hypogonadism due to any cause can cause a modestly increased final height due to late closure of epiphysis and prolonged growth of legs. Truncal growth, which is primarily mediated through sex hormones during adolescence, is impaired. Thus, these patients also have a short trunk (eunuchoid habitus).

c. **Disproportionate tall stature with intellectual deficit:** Included in this group are disorders involving duplication of X or Y chromosomes—the klinefelter syndrome (XXY, XXYY, XXXY and mosaics) and the XYY syndrome. Patients with Klinefelter syndrome are tall and thin with a low US : LS ratio. Genital abnormalities like small penis, hypospadias and cryptoorchidism may be present. Androgen function is initially preserved and almost all boys enter puberty. The testes are firm and small for the stage of pubertal development and testicular volume rarely exceeds 4–5 ml. However, further adolescent development may be delayed and oligospermia is usual. Most patients have some degree of intellectual impairment with increased prevalence of learning, behavioral and psychosocial problems.

Homocystinuria, an aminoaciduria, is associated with Marfanoid tall stature though prevalence of ocular complications like ectopia lentis and high myopia is higher and intellect is usually sub-normal.

(iii) Secondary causes of large size

Growth hormone excess, though rare, should be considered as a cause of tall stature and increased growth in all patients with these symptoms. The excess GH secretion is generally due to a functioning pituitary adenoma. If this occurs at an age when the epiphysis are open as in case of young children 'Gigantism' results. The height velocity increases and the growth curve progressively deviates from the normal. On the other hand, acromegaly is seen in adults whose epiphyses have fused. The adolescent who still has open epiphyses shows a mixed picture of rapid growth and acromegalic features. Patients often exhibit impaired vision, visual field defects and sometimes, features of elevated intra-cranial pressure. The serum GH levels are markedly elevated and are not suppressed by glucose administration during a standard glucose tolerance test. Neuro-imaging focusing on the hypothalamic-pituitary area is indicated for demonstrating the size and location of the adenoma before further neurosurgical management.

SUGGESTED READING

1. Allen DB, Cuttler L. Short stature in childhood: Challenges and choices. N Engl J Med 2013; 368:1220–8.

2. Backeljauw PF, Chernausek SD. The insulin like growth factors and growth disorders of childhood. Endocrinol Metab Clin N Amer, 2012;41:265–77.

3. Lee JM, Appugliese D, Coleman SM, et al. Short stature in a population based: Social, emotional and behavioural functioning. Pediatrics 2009;124:903–10.

4. Radivojevic U, Thibaud E, Samara-Boustani D, et al. Effects of growth reduction therapy using high-dose 17 beta-estradiol in 26 constitutionally tall girls. Clin Endocrinol (Oxf), 2006; 64:423–8.

5. Rozendaal L, le Cessie S, Wit JM, et al. The Dutch Marfan Working Group. Growth reductive therapy in children with Marfan syndrome. J Pediatr 2005;147:674–9.

6. Tanner JM. Fetus into man. Physical growth from conception to maturity. Harvard University Press, Cambridge, 1978.

7. Tatton-Brown K, Rahman N. Sotos syndrome. Eur J Hum Genet 2007;15:264–271.

4

Obesity and Thinness

KN Agarwal

OBESITY

Childhood obesity is a serious medical condition that affects children and adolescents. It occurs when a child is well above the normal weight for his or her age and height. Lifestyle issues—too little activity and too many calories from food and drinks—remain a significant contributor to childhood obesity.

Trends: The prevalence of childhood obesity is increasing world over. In USA it is monitored nationally by using data from the National Health and Nutrition Examination Survey (NHANES). The report

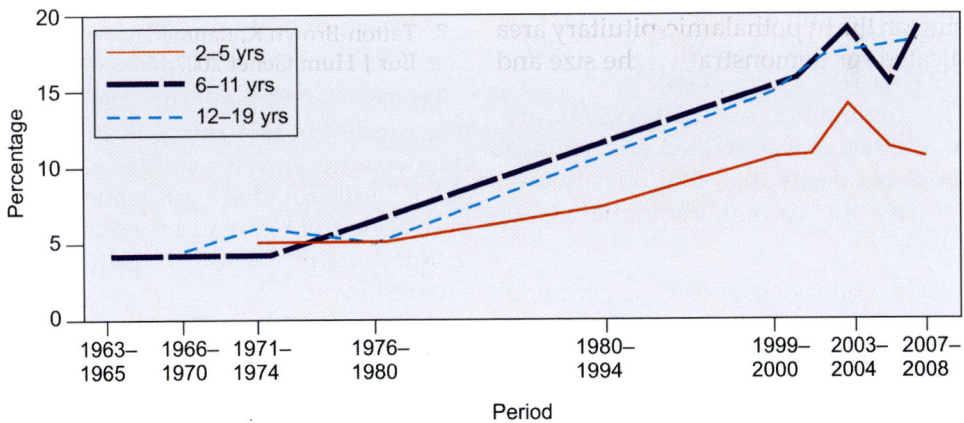

Data from the United States Health and Nutrition Examination Survey (NHANES). Obesity is defined as a body mass index (BMI) 95th percentile for age and gender. The figure does not distinguish between group with mild versus severe obesity.

Reproduced from: Centers for Disease Control and Prevention. CDC (Centre for Disease Control)[3] grand rounds: Childhood obesity in the United States. MMWR (Morb Mortal Wkly Rep) 2011;60:42.

Fig. 4.1: *Prevalence of obesity among children and adolescents, by age group—United States, 1963–2008*

presented by NHANES for the year 2007–2008 estimated that 16.9% of children and adolescent in the age group of 2–19 years were obese. Childhood obesity prevalence among preschool children between age group of 2–5-year-old girls and boys has increased from 5.0 to 10% between 1976–1980 and 2007–2008 and it has increased from 6.5 to 19.6% among age group of 6–11-year-old. The data (Ogden et al 2012)[1] collected for the same period shows that the adolescent (age group 12–19 years) obesity has increased from 5.0 to 18.1%.

The other interesting observations were by:

i. Madsen et al 2010,[2] reporting decline in overall obesity in California between 2001 and 2008, but with increase in black and American Indian girls.

ii. CDC during 2008–2011[3] recorded decrease in obesity in 19 of 43 states of USA, was stable in 20 and increased in 3 states.

iii. Claire et al 2011,[4] observed that the prevalence of severe obesity increased in three cycles of National Health and Nutrition Examination Survey (NHANES), USA from 1.2% (1976–1994) to 4.9% (1999–2004).

iv. The continuation of childhood obesity in adulthood was related to early age onset of obesity, parental obesity and severity of obesity (Cunningham et al 2014).[5]

Similarly, childhood obesity is common in the United Kingdom and according to the health survey conducted in 2004, obesity among 2–10 years old was 14% and among 11–15 years old was 15% (Stamatakis et al 2010).[6]

Recent Indian data of Khadilkar et al (2009)[7] showed obesity, as compared to Agarwal et al[8, 9] as increase in weight was from 20 to 29% among boys and 18 to 25% among girls without any significant increase in height.

The meta analysis of nine studies undertaken by Gedham 2013,[10] in which 92,862 subjects were identified and analyzed, the prevalence of overweight was estimated to be 12.64% (95% CI 8.48–16.80%) and that of obesity to be 3.39% (95% CI 2.58–4.21%).

Majority of Indian children are obese: Survey by *Yashika Kapoor*—August 25, 2010[11]

www.themedguru.com

- In 21 schools from cities of Delhi, Mumbai, Bengaluru, Chennai, Mangalore, Lucknow, Indore, Raipur, Coimbatore, Mohali, Baroda, Amritsar, Panipat and Moga.
- Investigators found that 43% of children, who were aged 7 and above, had more than the ideally required body mass index (BMI), an important pointer of overall fitness that is linked to obesity.

Definitions (Klish 2014)[12]

- 'Overweight' technically means an excess of body weight.
- 'Obesity' refers to excess of fat.

To assess 'overweight/obesity', body mass index (BMI) remains the accepted standard measure for children 2 years and older. BMI provides a guideline for weight in relation to height and is equal to weight in kg divided by the height in meters squared (m^2).

$$BMI = weight\ kg \div height\ m^2$$

Children are growing in weight as well as height; therefore the norms vary with age and sex. In adolescence BMI changes with sexual maturity. The BMI values in adults between 25 and 30 are for overweight; while obese have values 30 kg/m^2.

Table 4.1: *BMI for weight categories for children 2–18 years (international CDC and AAP)*

Category	Children[®]	Indian data
Underweight	BMI <5th percentile for age	*see* Tables 4.2 to 4.5
Normal	BMI ≥5th to <85th centile for age	
Overweight	BMI ≥85th <95th percentile	
Obesity	BMI ≥95th percentile	
Severe obesity	BMI ≥120% of the 95th centile	

[®] In children most accepted definition for 'Severe obesity' is ≥120% of the 95th centile or a BMI ≥35 (whichever is less). This corresponds to 99th percentile (Ref: Circulation 2013; p. 1689; AJCN 2009; p. 1314; Pediatrics 2012; p. 1136).

Circulation 2013; p. 1689—severe obesity in the USA is present in 4.7% of children 6–11 years and in 6.3% of the adolescents.

Table 4.2: *BMI mean ± SD and percentiles for Indian affluent boys*

Age (yrs)	N	Mean	SD	Percentiles						
				5	10	25	50	75	85	95
2	202	16.0	1.41	15.3	15.8	15.8	16.1	16.3	16.8	17.4
3	363	15.3	1.31	14.9	15.1	15.3	15.5	15.9	16.5	16.5
4	525	15.0	1.29	14.6	14.7	14.8	15.2	15.4	15.8	16.0
5	97	14.4	1.31	12.4	12.8	13.5	14.4	15.0	15.6	17.0
6	358	14.8	1.34	13.0	13.4	13.9	14.7	15.4	15.9	17.8
7	501	15.0	1.57	13.0	13.5	14.0	14.8	15.7	16.4	18.8
8	585	15.2	1.83	12.9	13.3	14.0	14.8	15.9	17.0	19.7
9	701	15.6	2.09	12.9	13.4	14.2	15.1	16.4	17.3	21.0
10	1135	16.1	2.42	13.2	13.6	14.5	15.4	17.0	18.5	22.1
11	1476	16.6	2.71	13.3	13.8	14.7	15.8	17.6	19.1	23.4
12	1652	17.1	2.72	13.6	14.2	15.2	16.4	18.3	19.8	23.8
13	1591	17.7	3.03	14.0	14.5	15.5	17.1	19.0	20.4	25.3
14	1433	18.2	2.90	14.5	15.1	16.3	17.7	19.6	21.1	25.3
15	1093	19.2	3.12	15.4	15.9	16.9	18.4	20.5	22.0	27.3
16	771	19.7	3.09	15.8	16.5	17.4	19.1	21.1	22.7	27.6
17	361	20.1	2.83	16.3	16.9	17.8	19.7	21.5	24.4	27.8
18	87	20.4	3.36	15.7	16.8	17.8	20.0	22.5	23.6	28.0

N: Number of children; BMI: Body mass index kg/m^2.
[Indian Pediatr 2001;38:1217–35].

Table 4.3: *BMI mean ± SD and percentiles for Indian affluent girls*

Age (yrs)	N	Mean	SD	Percentiles						
				5	10	25	50	75	85	95
2	179	16.1	1.41	15.4	15.4	15.8	16.1	16.3	16.4	16.6
3	266	15.6	1.39	14.8	14.9	15.0	15.5	15.8	16.0	16.3
4	432	15.2	1.45	14.3	14.5	14.7	15.2	15.6	15.9	16.1
5	254	14.4	1.5	12.3	12.7	13.5	14.3	15.2	15.7	18.3
6	449	14.5	1.7	12.4	12.9	13.5	14.3	15.3	16.0	18.8
7	596	15.0	1.9	12.5	12.9	13.5	14.6	15.7	16.6	19.7
8	640	15.7	2.3	12.8	13.2	13.9	14.9	16.5	18.0	21.4
9	784	15.7	2.5	12.5	13.1	14.0	15.1	16.8	18.0	21.7
10	933	16.7	6.6	13.0	13.6	14.6	16.1	18.2	19.9	23.2
11	906	17.5	3.1	13.5	14.1	15.2	16.9	19.0	20.6	24.5
12	893	18.4	3.2	13.9	14.6	15.9	17.8	20.1	21.9	25.7
13	782	19.2	3.6	14.6	15.3	16.7	18.6	21.0	22.6	27.1
14	627	19.7	3.2	15.4	16.4	17.3	19.0	21.4	23.0	27.4
15	383	20.0	3.3	15.9	16.5	17.7	19.3	22.0	23.6	27.7
16	270	20.5	3.2	15.9	16.6	18.1	20.0	22.4	23.7	27.4
17	119	20.3	3.1	16.6	16.9	18.3	20.1	22.0	23.0	25.9
18	27	20.9	3.2	16.9	17.9	18.3	20.0	23.0	23.2	

N: Number of children
[Indian Pediatr 2001;38:1217–35].

Table 4.4: *Showing body mass index (BMI) in adolescents in relation to sexual development and age*

Age (yrs)		Percentiles										
		5	15	50	85	90	95	5	15	50	85	90	95
		SMR-2						SMR-3					
9	G	13.4	14.2	16.5	20.1	20.7	21.7	—	—	—	—	—	—
	B	13.0	13.7	15.4	17.6	17.9	18.7						
10	G	13.7	14.6	16.7	20.2	21.0	22.6	14.7	15.6	18.0	21.5	22.6	23.8
	B	13.5	14.3	15.9	18.7	20.0	21.5	13.7	14.7	16.1	18.9	19.4	20.1
11	G	13.3	14.2	15.8	18.9	20.2	22.3	14.9	15.8	18.0	21.2	22.2	24.1
	B	13.5	14.4	15.9	19.1	20.2	21.6	14.1	15.0	16.6	19.0	19.8	21.3
12	G	13.4	14.2	15.8	18.9	19.5	20.8	14.7	15.6	17.6	20.8	21.9	24.3
	B	13.6	14.4	16.2	19.5	20.8	22.4	14.2	15.2	16.7	19.8	20.9	23.5
13	G	13.3	14.2	15.4	18.3	19.6	21.4	14.5	15.3	17.4	20.1	21.4	23.3
	B	14.0	14.7	16.4	21.1	22.3	24.5	14.0	14.9	16.8	20.2	20.7	23.2
14	G	—	—	—	—	—	—	15.1	15.7	17.9	21.0	21.8	23.2
	B	—	—	—	—	—	—	—	—	—	—	—	—

Contd.

Table 4.4: *Showing body mass index (BMI) in adolescents in relation to sexual development and age (Contd.)*

Age (yrs)		Percentiles										
		5	15	50	85	90	95	5	15	50	85	90	95
				SMR-4						SMR-5			
11	G	16.1	16.7	19.0	23.0	23.6	24.9	—	—	—	—	—	—
	B	—	—	—	—	—	—						
12	G	15.7	16.9	19.2	22.1	23.0	24.3	17.0	19.2	22.5	25.7	26.9	27.8
	B	14.5	15.6	17.0	19.8	20.9	22.3	—	—	—	—	—	—
13	G	15.6	16.6	18.9	21.8	22.6	24.9	17.0	18.6	21.6	25.2	26.3	27.8
	B	14.9	15.7	17.8	21.1	21.9	44.4	15.5	16.9	18.7	21.7	22.6	23.7
14	G	15.5	16.5	18.8	21.9	23.0	24.3	16.8	18.4	21.2	24.5	26.2	27.7
	B	14.7	15.8	17.7	20.8	21.7	23.8	15.4	16.4	18.6	22.3	23.5	25.2
15	G	16.0	16.9	19.0	22.3	23.5	25.2	16.5	17.8	20.7	25.1	25.8	26.7
	B	15.4	16.4	18.4	22.0	23.7	25.7	15.5	16.5	18.9	22.3	23.9	25.8
16	G	16.1	17.2	19.6	23.1	24.5	25.9	16.5	17.5	21.2	24.9	26.7	28.1
	B	15.9	16.9	18.9	22.7	23.9	25.3	15.8	17.1	19.3	22.7	23.8	26.0
17	G	16.5	17.1	19.5	21.6	22.1	23.2	16.7	18.2	21.3	24.2	25.3	26.3
	B	16.2	17.1	19.3	22.3	23.4	25.5	16.2	17.3	20.1	23.0	23.9	26.2

SMR: Sexual maturity rating; G: Girls and B: Boys.
Indian Pediatr 2001;38:1217–35

BMI—easy to measure

- Inexpensive
- Standardized cut off points for overweight and obesity: Normal weight is a BMI between 18.5 and 24.9; overweight is a BMI between 25.0 and 29.9; obesity is a BMI of 30.0 or higher. Strongly correlated with body fat levels, as measured by the most accurate methods.
- Hundreds of studies show that a high BMI predicts higher risk of chronic disease and early death.

BMI—limitations

- Indirect and imperfect measurement—does not distinguish between body fat and lean body mass.
- Not as accurate a predictor of body fat in the elderly as it is in younger and middle-aged adults.

Table 4.5: *BMI values to define overweight and obesity among Indian adolescent in relation to SMR staging (Indian affluent children data)*

SMR	Sex	Overweight (>85th centile)	Obese (>95th centile)
2	Boys	19	22
	Girls	19	21
3	Boys	19	20
	Girls	21	23
4	Boys	22	25
	Girls	22	24
5	Boys	22	26
	Girls	25	27

- At the same BMI in women has, on average, more body fat than men, and Asians have more body fat than white.

BMI centile curves for Indian affluent children are given in Figs 1.13 and 1.14; Chapter 1. BMI values from WHO-NCHS are given in Figs 4.2 and 4.3; Appendix Figs 1 and 2 and Table 4.5.

Skin fold thickness (SFT) is a useful tool to evaluate the regional fat distribution, values > 95th centile for age and sex being used to identify children with obesity. Mean of triceps and biceps SFT represents peripheral obesity, while mean of subscapular and suprailiac SFT represent central obesity (Appendix Tables 20 and 21).

Trend in Obesity

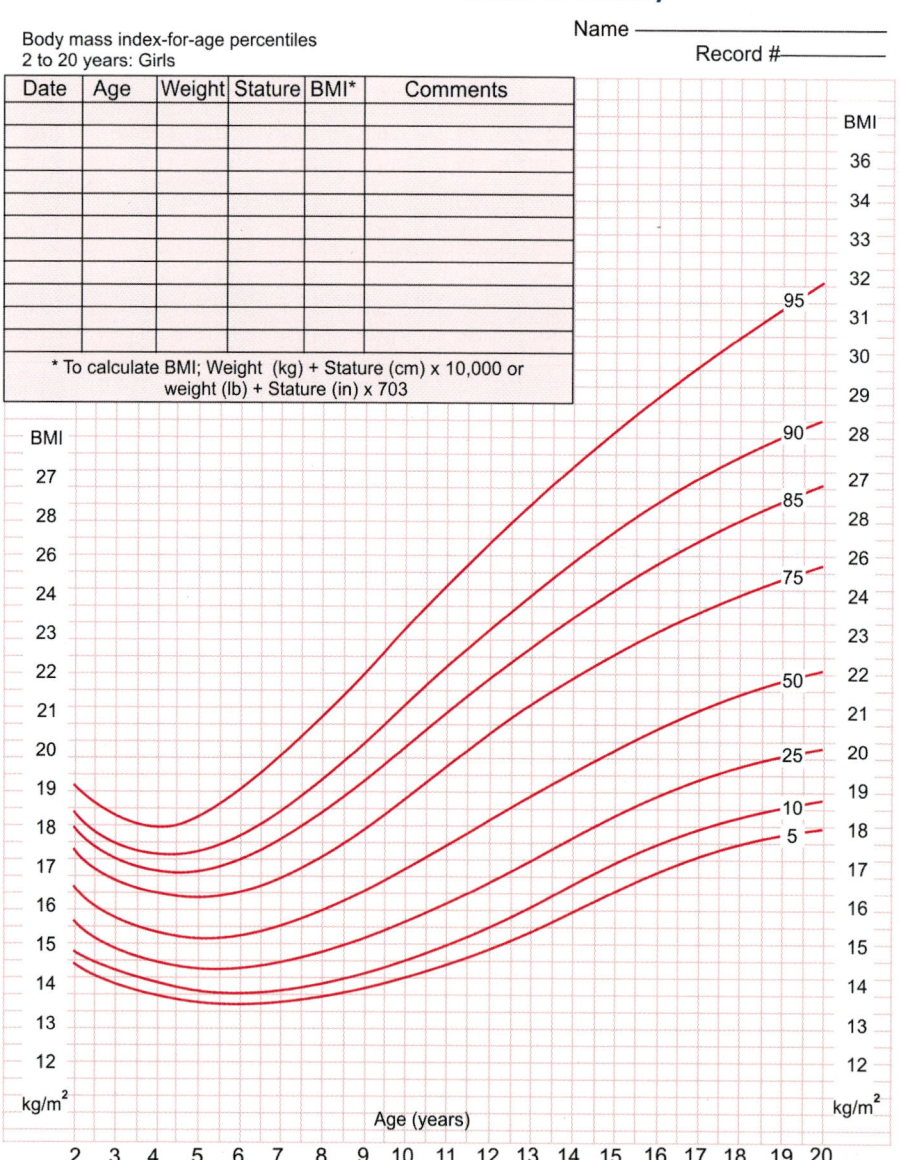

Fig. 4.2: *BMI for age centiles for girls (CDC 2000)*

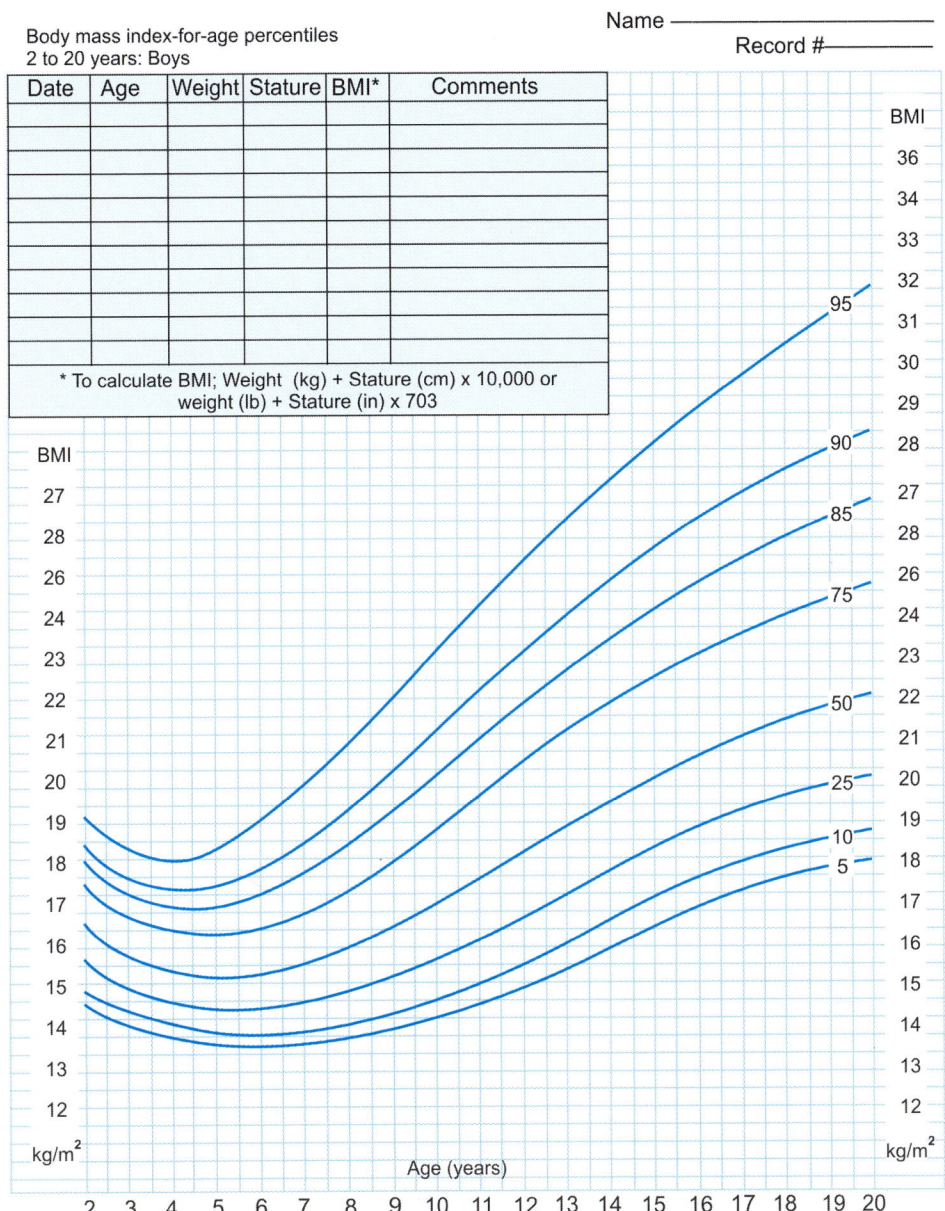

Body mass index-for-age percentiles
2 to 20 years: Boys

Name ——————————

Record #————————

Date	Age	Weight	Stature	BMI*	Comments

* To calculate BMI; Weight (kg) + Stature (cm) x 10,000 or
weight (lb) + Stature (in) x 703

Fig. 4.3: *BMI for age centiles for boys (CDC 2000)*

Waist Circumference

Waist circumference is the simplest and most common way to measure "abdominal obesity"—the extra fat found around the middle that is an important factor in health, even independent of BMI. It is the circumference of the abdomen, measured at the natural waist (in between the lowest rib and the top of the hip bone), the umbilicus (belly button), or at the narrowest point of the midsection.

Strength

- Easy to measure
- Inexpensive
- Strongly correlated with body fat in adults as measured by the most accurate methods
- Studies show waist circumference predicts development of disease and death.

Limitations

- Measurement procedure has not been standardized
- Lack of good comparison standards (reference data) for waist circumference in children.
- May be difficult to measure and less accurate in individuals with a BMI of 35 or higher.

Waist Circumference Indices

Waist-to-Hip Ratio (***Report of a WHO Expert Consultation, Geneva, 8–11 Dec 2008***) Like the waist circumference, the waist-to-hip ratio (WHR) is also used to measure abdominal obesity. It is calculated by measuring the waist and the hip (at the widest diameter of the buttocks), and then dividing the waist measurement by the hip measurement.

Strength

- Good correlation with body fat as measured by the most accurate methods
- Inexpensive
- Studies show waist-to-hip ratio predicts development of disease and death in adults.

Limitations

- More prone to measurement error because it requires two measurements
- More difficult to measure hip than it is to measure waist
- More complex to interpret than waist circumference, since increased waist-to-hip ratio can be caused by increased abdominal fat or decrease in lean muscle mass around the hips
- Turning the measurements into a ratio leads to a loss of information: Two people with very different BMIs could have the same WHR
- May be difficult to measure and less accurate in individuals with a BMI of 35 or higher.

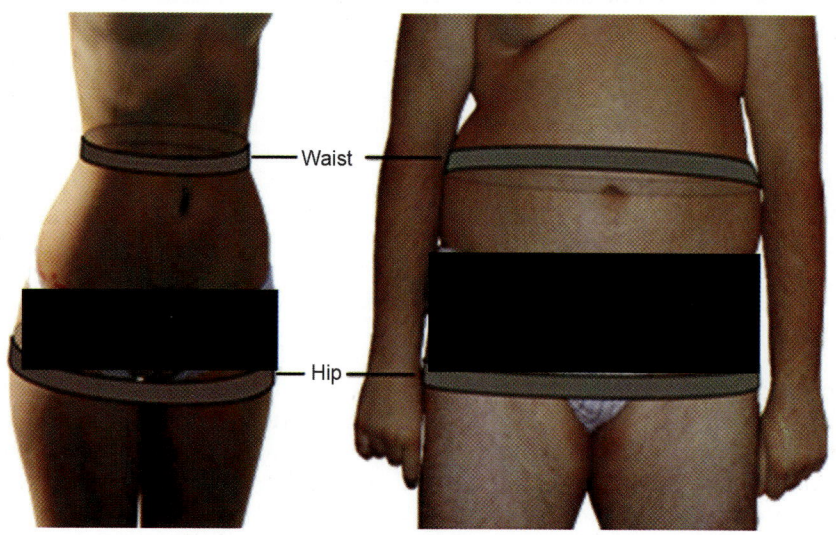

Fig. 4.4: *Waist, hip circumferences in girls*

WHR (waist/hip ratio)— an index of obesity

- A WHR of 0.7 for women and 0.9 for men have been shown to correlate strongly with general health and fertility.

- Women within the 0.7 range have optimal levels of estrogen and are less susceptible to major diseases such as diabetes, cardio-vascular disorders and ovarian cancers.

- Men with WHRs around 0.9, similarly, have been shown to be more healthy and fertile with less prostate cancer and testicular cancer.

Bogalusa Heart Study reported in the October 11 issue of BMC Pediatrics—waist-to-height ratio (WHtR)

- In the normal-weight group, 9.2% of the children were centrally obese (WHtR ≥0.5), whereas 19.8% of the overweight/obese group were not (WHtR <0.5).
- Significantly adverse levels of low density lipoprotein (LDL) cholesterol, triglycerides, and insulin were more likely to occur in normal weight, centrally obese children (1.66, 2.01, 1.47 and 2.05 times more likely, res-pectively), based on multivariate analysis.
- The normal weight, central obesity group had a higher prevalence of a parental history of type 2 diabetes mellitus and a significantly higher prevalence of metabolic syndrome ($P <.0001$).

Waist-to-height ratio (WHtR) may pre-dict cardiometabolic risk in normal-weight as well as in overweight/obese children, according to results from the Bogalusa Heart Study reported in the October 11 issue of *BMC Pediatrics*.

The study sample consisted of 3091 children aged 4–18 years; half were boys, and 56% were white. The investigators determined cross-sectional cardiometabolic risk factor variables and used age-, race-, and sex-specific BMI percentiles to classify children as normal weight (5th–85th per-centiles) or overweight/obese (e 85th percentile). On the basis of the WHtR), the investigators compared the cardiovascular risk profiles of each group. "The findings of the present study emphasize the utility of WHtR not only in detecting central intra-abdominal obesity and related cardio-metabolic risk among normal weight children, but also in identifying those without central obesity and a healthy risk factor profiles among the overweight/obese children," the study authors write. "Thus, WHtR has a potential for wider use as a simple measure to assess cardiometabolic risk in pediatric primary care practice." *BMC Pediatr* 2010;10:73.

Etiology/Pathogenesis

1. **Environmental factors:** Obesity in chil-dren is strongly influenced by environ-mental factors, caused by sedentary lifestyle or a caloric intake that is greater than need, i.e. overeating/overfeeding.

i. Increasing tendency to consume higher glycemic index food, sugar containing beverages (including fruit juices), large portions of prepared foods (combo choices), fast food service, school meal nutrition content

ii. Diminishing family presence at meals

iii. Decreasing structured physical activity

iv. Increasing use of computer oriented play activity (this displacement of phy-sical activity/depression of metabolic rate/adverse effect on diet quality), this time spent is directly related to obesity in children and adolescents. Reducing TV viewing decreases obesity.

v. Video games have weaker influence on obesity as compared to TV. Some of the nutrition education and physical activity games cause moderate increase in energy expenditure also.

iv Non-availability of playgrounds and sidewalks

"These are controllable factors".

2. **Sleep:** Shortened sleep duration is asso-ciated with obesity or insulin resistance.

Studies have shown that "Sleep fragmentation and intermittent hypoxemia" caused by sleep disordered breathing are associated with decreased insulin sensitivity in adolescents, independent of adiposity. This may be causing alterations in serum leptin and ghrelin levels, both are implicated in regulation of appetite, or perhaps a longer opportunity to ingest food.

3. **Medications:** Drugs cause weight gain
   a. Psychoactive drugs—olanzapine and risperidone
   b. Antiepileptic drugs—carbamazepine, valproate
   c. Glucocorticoids—however brief course of inhaled steroids does not increase weight.

4. **Virus:** There is substantial proof that adiposity can be triggered or exacerbated by exposure to a virus named Adenovirus 36.

5. **Toxins:** Endocrine disrupting chemicals such as bisphenol A (BPA), used to manufacture polycarbonate resin. A common contaminant of foods sold in cans and plastic packaging is a selective modulator of estrogen receptors, and accelerates adipogenecity and postnatal somatic growth. In adults and children its urinary concentration is associated with obesity and related diseases including diabetes and hypertension.

6. **Genetic factors:** Play a definite role in interaction with environmental factors to produce obesity. The genetic factors are responsible for 30–50% variation in adiposity.

A variety of specific syndromes and single gene defects amounting to around 1% are identified. These children along with overweight and obesity have dysmorphic facies, mental retardation, short stature, retinal changes or deafness. For examples, Prader-Willi syndrome. A single gene defect regulating body weight (severe obesity), pigmentation having mutations in the melanocortin 4 receptor and the reproductive system involvement is described.

Table 4.6: *Genetic causes of obesity*

Single gene disorders	Chromosomes	Clinical
Leptin deficiency	7q31.3	Hypometabolic rate, hyperphagia, pubertal delay, impaired glucose tolerance, hypothalamic hypogonadism
Pro-opiomelanocortin deficiency (POMC)	2p23.3	Severe, early onset obesity, red hair, hyperphagia, adrenal insufficiency, hyperpigmentation
Proprotein convertase 1 (PCSK1, as known as prohormone convertase 1)	5q15–q21	Early onset obesity, abnormal glucose homeostasis, hypogonadotrophic hypogonadism, hypocortisolism, elevated plasma proinsulin and POMC
Melanocortin receptor 4 haploinsufficiency (MC4R)	18q21.3–q22	Early onset moderate-severe obesity and hyperphagia, increased bone density
Leptin receptor deficiency (LEPR)	1p31–p22	Severe early onset obesity, hypometabolic rate, hyperphagia, pubertal delay, hypothalamic hypogonadism

Endocrine disease: Identified in <1% children and adolescents with obesity, i.e. hypothyroidism, cortisol excess Cushing's syndrome corticosteroid medication. Cushing syndrome, growth hormone deficiency, and pseudohypoparathyroidism are associated with mild obesity/overweight. Acquired hypothalamic lesions after surgery may cause severe obesity.

Metabolic programming: There is evidence to suggest that environmental and nutritional factors during critical periods in development can have permanent effects on individual's predisposition to obesity and metabolic diseases.

Maternal nutrition or endocrine profile: During gestation is probably an important determinant of metabolic programming:

a. Births small for gestation (SGA), large for gestation (LGA) or preterm have higher rates of insulin resistance, later life diabetes, hypertension, heart disease and obesity.

b. Dutch famine studies 1944–1945 also showed association between nutritional exposures during gestation and later obesity and metabolic diseases.

c. Mothers prepregnancy weight and weight gain in pregnancy are important predictors of the infant birth weight.

d. In a Swedish study maternal diabetes was associated with increased BMI in male offspring in adult life.

e. Children of gastric bypass surgery mothers appear to have lower prevalence of obesity than those born before bypass. Thus, reversal of maternal obesity had beneficial permanent effects on the metabolic profile of the offspring.

f. Faster rate of weight gain in infancy or early childhood may develop obesity or metabolic syndrome in childhood.

Other maternal endocrine factors: Younger age of mother at menarche was an independent predictor of the child's obesity. Their children also had more rapid growth during the first 2 years of life.

Complications

Obesity in children and adolescents affects many organ systems. Table 4.7 lists the various complications of obesity.

Obesity is a major risk factor for cardiovascular disease in the pediatric age group, the other risk factors being decreased physical activity, dyslipidemia, hypertension and tobacco use. Since obesity predisposes to all of these except tobacco use, obese children are at a substantial risk of suffering from cardiovascular disease. They are found to have a 2.4 fold risk for increased blood pressure and up to 10 fold risk for adult hypertension, a greater increase being noted in children with abdominal adiposity. In a study from Ludhiana 15.33% of overweight and 93.1% of obese children were found to have hypertension (Mohan et al 2004).

The prevalence of diabetes mellitus is also high in obese children and adolescents. A global increase in prevalence of type II DM is noted parallel to increase in the prevalence of obesity. Thus, Japan has reported a 30 fold increase in its prevalence over the

---

Table 4.7: *Complications of obesity*

- Cardiovascular disease
- Hypertension
- Deranged lipid profile
- Steatohepatitis, gallstones
- *Endocrine consequences:* Insulin resistance, Type 2 DM, and hyperandrogenism
- Menstrual abnormalities, polycystic ovaries
- *Orthopedic sequalae:* Genu varum, valgus deformity, slipped capital femoral epiphysis, tibia vara
- Pseudotumor cerebri
- *Psycho-social consequences:* Poor self-esteem, depression, poor scholastic performance

last 20 years. A high prevalence of insulin resistance and hyperinsulinemia has also been reported from post-pubertal urban Indian children in association with abdominal obesity and excess body fat (Seth Sharma 2013).[13]

Apart from the various medical complications, obesity leads to many psychosocial problems too. An obese child often gets discriminated against by family and friends. They may be teased or ridiculed as being clumsy, unattractive and overindulgent. This leads to a lowered self-image in obese children and withdrawal from social contact.

Evaluation

A child with obesity needs a comprehensive medical and nutritional evaluation. The purpose is to:

1. Ascertain the etiology: Whether the obesity is 'simple' or 'pathological' and
2. Screen for the complications related to obesity.

Table 4.8 summarizes the clinical evaluation of an obese child.

A child with simple obesity is usually tall for age/genetic potential, though weight-for-height is disproportionately higher. On the other hand, most children with pathological obesity are short. One should look for thyroid enlargement, clinical features of Cushing syndrome and other genetic syndromes associated with obesity. Most of the genetic syndromes associated with obesity are characterized by short stature, developmental delay, dysmorphic features and hypogonadism. Coincidence of several of these features should prompt a work up for syndromic obesity. A child with normal cognitive function and good linear growth is very unlikely to have syndromic obesity. As mentioned earlier, children with hypothyroidism and GH deficiency have increased weight-for-height though they are seldom grossly obese. Obese pre-pubertal boys often seem to have micropenis, many

Table 4.8: *Clinical evaluation of an obese child*

History

Dietary details
Activity pattern
Mental development
School performance
Drug intake
H/o CNS injury/infection
Snoring, day time sleep
Psychosocial concerns
Prevalence of obesity and body fat distribution in other family members
Prevalence of obesity related morbidity in family
Review of past weight and height records, if available

Examination

Height, weight: BMI calculation
Parental height
Blood pressure
Acanthosis nigricans
Pubertal staging
Dysmorphic features
Development assessment

Table 4.9: *Investigations of an obese child*

Suspected pathological obesity

Condition suspected	Test
Hypothyroidism	FT_4, TSH
Cushing Syndrome	Urine free cortisol or morning and evening plasma cortisol
GH deficiency	GH provocative testing
Prader-Willi syndrome	Karyotyping; deletion of chromosome 15
CNS disorder	MRI brain

Consequences of obesity

- Fasting lipid profile
- Fasting and PP blood sugar
- AST, ALT—If elevated: Consider USG abdomen for steatohepatitis

of them may even present with it. However, once the pubic fat is pushed back and penile length measured from pubic symphysis to tip of glans penis, the length is found to be normal for age.

All children with BMI > 85th centile should be investigated as per Table 4.5. Bone age should be assessed in all obese children. It is mildly advanced, but appropriate for height age in children with simple obesity while children with hypothyroidism and GH deficiency would have a retarded bone age.

Management

For pathological obesity, treatment of underlying condition is indicated where applicable and possible. For simple obesity, the corner stones of management are dietary management, a regular exercise program and lifestyle/behavior modification. Pharmacological interventions have limited role in management of childhood obesity.

The best time to intervene for obesity prevention is the pre-school period. This would not only limit the proliferation of adipose cells, but also, more importantly, inculcate healthy dietary and lifestyle practices in the child.

Table 4.10 summarizes the important principles for obesity prevention at different ages.

Role of Physicians in Management of Childhood Obesity

Physicians need to

a. Identify overweight children by recording height, weight and BMI at visits. Many parents do not recognize the overweight and BMI problem of their child.

b. Screen obese children for obesity related complications.

c. Promote a healthy lifestyle and be a role model for the same for the children under their care and their families.

Table 4.10: *Universal anticipatory guidance for obesity prevention*

Age group	Lifestyle targets
Infants	• Encourage sustained breastfeeding (>3–6 months) • Discourage early introduction of solid foods before 4 months age • Goal: *Moderate* rates of weight gain, including in low-birth weight infants (throughout childhood). Rapid catch-up growth may be detrimental.

(contd.)

Table 4.10: *Universal anticipatory guidance for obesity prevention (contd.)*

Age group	Lifestyle targets

Toddlers	• Nutritional:
	  – Continue to broaden diet, emphasize vegetables, fruits
	  – Minimize intake of juice and other sweetened beverages
	• Physical activity/*yoga*:
	  – Establish habits of physical activity (playground, outdoor time)
	  – Establish healthy television habits (<1 hour/day; not at meals, minimize number of televisions in household)
	• Behavioral:
	  – Emphasize family-based means, avoid cooking special meals for kids
	  – Do not use food as a reward or punishment
	  – Do not encourage eating beyond satiety (no "clean plate club")
	  – Provide parental modeling of healthy diet (emphasizing vegetables), physical activity, and minimal television viewing
	  – Offer positive reinforcement for healthy choices, avoid criticism

School-age	*All of the above plus:* Children
	• Physical activity/*yoga*:
	  – Investigate local opportunities for adding organized sports to lifestyle (town programs, YMCA, school). Goal: At least 1 structured activity every season.
	  – Offer options, including individual sports if team sports not practical or enjoyed by child (martial arts, dance, swimming)
	  – Participate in physical activities with children: Recreational sports, outdoor play, walking, or bicycling
	• Behavioral:
	  – Support healthy body image, emphasizing strength and health rather than weight and appearance

Adolescents	Watch out for and *discourage:*
	• Nutritional:
	  – Excessive take-out or restaurant meals
	  – Meal skipping or inadequate meals (which often lead to out-of control eating later in the day)
	  – 'Grazing' rather than meal-based eating habits
	  – Withdrawing from sports or other physical activity

Thinness

There are several factors to take into consideration when evaluating your child's weight. Has he always been thin? Are both of his parents very thin? A child who has a genetic tendency to be thin is in a different boat than a child who has always been normal to hefty and who has recently stopped gaining—or started losing—weight. Even if your child has just recently thinned out, though, there may be nothing to worry about. When an increase in height precedes a gain in weight, your child may appear underweight for a while, until his weight gain catches up.

Gomez et al 1956[14] introduced malnutrition classification of weight below a specified percentage of median weight for the child's age. Later Seoane and Latham 1971[15] proposed splitting weight-for-age into weight-for-height and height-for-age, allowing underweight to be defined as wasting or stunting, or both. Subsequently Waterlow et al 1972[16] recommended the use of Z-scores for the definitions of under-weight, wasting, and stunting, with the cut-offs defined in terms of standard deviations (SDs) below the median rather than as percentages of the median *see* Chapter 12.

In 1983, WHO formally recognized the US National Center for Health Statistics (NCHS) classification as the international reference and has used it since to classify children as underweight, wasted, or stunted, each based on a cut-off of –2Z-scores. Wasting in Particular is assessed with the NCHS/WHO weight-for-height reference, which compares the child's weight to the average weight of children of the same height. This ignores the child's age, which allows nutritional status to be assessed when age is not known. It also assumes that, on average, children of a given height, weight the same whatever their age; in infancy and adolescence, however, the weight–height relation depends on age.

Indices based on Weight and Height Measures

In children the three most commonly used anthropometric indices to assess their growth status are:

  i. Weight-for-height,
  ii. Height-for-age, and
  iii. Weight-for-age.

These anthropometric indices can be interpreted as follows:

Low weight-for-height: Wasting or thinness indicates in most cases a recent and severe process of weight loss, which is often asso-ciated with acute starvation and/or severe disease. However, wasting may also be the result of a chronic illness or psychological disorder. Provided there is no severe food shortage, the prevalence of wasting is usually below 5%, even in poor countries. In India where higher prevalence is found, is an important exception. On the severity index, prevalence between 10% and 14% is regarded as serious, and above or equal 15%

as critical. Lack of evidence of wasting in a population does not imply the absence of current nutritional problems: Stunting and other deficits may be present.

High weight-for-height: 'Overweight' is the preferred term for describing high weight-for-height. Even though there is a strong correlation between high weight-for-height and obesity as measured by adiposity, greater lean body mass can also contribute to high weight-for-height. On an individual basis, therefore, 'fatness' or 'obesity' should not be used to describe high weight-for-height. However, on a population-wide basis, high weight-for-height can be con-sidered as an adequate indicator of obesity.

Low height-for-age: Stunted growth reflects a process of failure to reach linear growth potential as a result of suboptimal health and/or nutritional conditions. On a popu-lation basis, high levels of stunting are associated with poor socioeconomic con-ditions.

Low weight-for-age: Weight-for-age reflects body mass relative to chronological age. It is influenced by both the height of the child (height-for-age) and his or her weight (weight-for-height), and its composite nature makes interpretation complex. For example, weight-for-age fails to distinguish between short children of adequate body weight and tall, thin children.

The Z-score or Standard Deviation Classification System

There are three different systems by which a child or a group of children can be compared to the reference population: Z-scores (standard deviation scores), percentiles, and percent of median. For population-based assessment—including surveys and nutritional surveillance—the Z-score is widely recognized as the best system for analysis and presentation of anthropometric data because of its advan-tages compared to the other methods.

The Z-score system expresses the anthropometric value as a number of standard deviations or Z-scores below or above the reference mean or median value. A fixed Z-score interval implies a fixed height or weight difference for children of a given age. For population-based uses, a major advantage is that a group of Z-scores can be subjected to summary statistics such as the mean and standard deviation. The formula for calculating the Z-score is:

Z-score (or SD-score) = (observed value – median value of the reference population)/standard deviation value of reference population

Interpreting the results in terms of Z-scores have several advantages:

- The Z-score scale is linear, and therefore, a fixed interval of Z-scores has a fixed height difference in cm, or weight difference in kg, for all children of the same age. For example, on the height-for-age distribution for a 36-month-old boy, the distance from a Z-score of –2 to a Z-score of –1 is 3.8 cm. The same difference is found between a Z-score of 0 and a Z-score of +1 on the same distribution. In other words, Z-scores have the same statistical relation to the distribution of the reference around the mean at all ages, which makes results comparable across age groups and indicators.
- Z-scores are also sex-independent, thus permitting the evaluation of children's growth status by combining sex and age groups.
- These characteristics of Z-scores allow further computation of summary statistics such as means, standard deviations, and standard error to classify a population's growth status.

Prevalence-based Reporting

For consistency with clinical screening, prevalence-based data are commonly reported using a cut-off value, often <–2 and >+2 Z-scores. The rationale for this is the statistical definition of the central 95% of a distribution as the 'normal' range, which is not necessarily based on the optimal point for predicting functional outcomes.

The WHO Global Database on Child Growth and Malnutrition uses a Z-score cut-off point of <–2SD to classify low weight-for-age, low height-for-age and low weight-for-height as moderate and severe undernutrition, and <–3SD to define severe undernutrition. The cutoff point of >+2SD classifies high weight-for-height as overweight in children.

The use of –2 Z-scores as a cut-off implies that 2.3% of the reference population will be classified as malnourished even if they are truly 'healthy' individuals with no growth impairment. Hence, 2.3% can be regarded as the baseline or expected prevalence. To be precise the reported values in the surveys would need to subtract this baseline value in order to calculate the prevalence above normal. It is important to note, however, that the 2.3% figure is customarily not subtracted from the observed value. In reporting underweight and stunting rates this is not a serious problem because prevalence in deprived populations is usually much higher than 2.3%. However, for wasting, with much lower prevalence levels, not subtracting this baseline level undoubtedly affects the interpretation of findings.

The World Health Organization defines grade 2 thinness in adults as BMI <17. Cole et al[17] after a multinational survey (2007) suggested, that a BMI of 17 at age 18 is a suitable cut-off to use as the basis for an International definition of thinness in children and adolescents. Three different criteria lead to this conclusion: BMI 17 is the WHO grade 2 cut-off for thinness in adults; BMI 17 at age 18 corresponds to a mean Z-score of –2, and, BMI 17 at age 18 is 80% of the median. The latter two criteria mean that in childhood the new cut off will be similar in Z-score and % of the median terms to those used before, notably the WHO definition of wasting—that is, weight-for-height below –2SD or 80% of the median.

For children 2–18 years CDC, the American Academy of Pediatrics, Institute of Medicine Endocrine Society and International Obesity. Task Force suggested BMI <5th percentile for age and sex as underweight. This seems more appropriate as recent studies by Natale and Rajgopalan 2014,[18] have concluded that local country data are more appropriate than using the WHO 2006[19] or NCHS standards. The Indian children BMI for 2–18 years in 5th percentile remains between 12.4 to 16.3 for boys and 12.3 to 16.9 for girls (2–17 years of age). Further in adolescence the table for BMI in relation to sexual maturity rating be used see table below (Table 4.11).

Thus, using Cole et al[17] criteria of BMI <17.0 at 18 yr will not be suitable for Indian children (Table 4.12).

Table 4.11: *BMI cut-off to define thinness in adolescents in different stages of sexual development (Indian affluent children data)*

SMR stage	BMI <5th centile
2	13.5 (boys, girls)
3	Boys <14, girls <15
4	Boys <14, girls <16
5	Boys <15.5, girls <16.5

Evaluation of a Thin Child

In our country tuberculosis, urinary tract infection and faulty weaning/feeding remain very common. After confirming thinness. After confirming presence of thinness, a clinician needs to examine children carefully to exclude any underlying chronic systemic disorder. Physical examination should include a detailed anthropometry including length/height, weight, head circumference, US:LS ratio, skin told thickness and mid-arm circumference. Signs of vitamin and nutrient deficiencies should be looked for along with a thorough systemic examination. A review of previous weight and height record if available, is invaluable. In some cases it may be worth while to monitor a child longitudinally for weight gain and height velocity.

A child with a documented weight loss, one showing a fall-off from the previously established growth curve centile or one who is stunted along with being thin need a more meticulous evaluation due to a higher possibility of an underlying organic cause. Important causes of thinness are listed in Table 4.12. The most important cause of thinness is malnutrition. According to NFHS II (1998–99); 47% of children in the age group of 6 months– 3 years are underweight and 46% are stunted. Around 60% of under 5 years of age deaths in India are due to diarrhea, measles, pneumonia and malaria these have associated malnutrition.

Table 4.12: *Causes of thinness*

1. Malnutrition
2. Systemic disorders
   - Infections/infestations
   - Malabsorption
   - Renal failure
   - Renal tubular acidosis
   - Asthma
   - Congenital heart disease
   - Cystic fibrosis
   - Diabetes mellitus
   - Malignancies
3. Altered growth potential
   - Prenatal insult
   - Chromosomal abnormalities

Laboratory Investigations

In most instances a detailed examination excludes an underlying organic cause for thinness. In cases with suspected pathological thinness, initial investigations include a complete blood count, urine and stool examination, urine culture, tuberculin test, blood urea, serum creatinine and LFT including serum proteins. Bone age estimation may be needed for some cases. More invasive diagnostic procedures are called for when a specific diagnosis is suspected.

Management

A child with an underlying organic cause needs to be treated for the same. A thin child with normal height, appetite and activity often needs follow-up after routine deworming and age-appropriate nutritional support.

If good height/weight gain velocity is observed during follow-up, no further management is called for. However, if appetite remains poor or weight and height gain falter, a re-evaluation is required.

NFHS-3 (INDIA)

Definitions—*Used in NFHS-3 survey (Fig. 4.5)*

1. A **stunted** child has a height-for-age Z-score that is at least 2 standard deviations (SD) below the median for the WHO Child Growth Standards. Chronic malnutrition is an indicator of linear growth retardation that results from failure to receive adequate nutrition over a long period and may be exacerbated by recurrent and chronic illness.
2. A **wasted** child has a weight-for-height Z-score that is at least 2SD below the median for the WHO Child Growth Standards. Wasting represents a recent failure to receive adequate nutrition and may be affected by recent episodes of diarrhea and other acute illnesses.
3. An **underweight** child has a weight-for-age Z-score that is at least 2SD below the median for the WHO Child Growth Standards. This condition can result from either chronic or acute malnutrition, or both.

The use of WHO 2006 data could have calculated higher values for stunting, wasting and underweight as compared to Indian children growth data.[8, 9]

- Indicators for SAM (severe acute malnutrition) a child with height-for-age standard deviation less than 3 and/or bilateral pitting edema. If one of these signs is detected, the child is suffering from SAM.

Calculations: For weight-for-age and weight-for-height.

Weight-for-age

$$= \frac{\text{Weight of the child}}{\substack{\text{Weight of the reference child} \\ \text{of the same age}}} \times 100$$

Weight-for-height

$$= \frac{\text{Weight of the child}}{\substack{\text{Weight of the reference child} \\ \text{of the same age}}} \times 100$$

WHO *indicators of underweight and malnutrition derived from the weight and height of children relative to their age*		
Index	**Cut-off value based on standard deviation (SD)/ percentage**	**What it indicates**
Weight-for-age	Less than –2 and more than –3	Moderate underweight
Weight-for-age	Less than –3	Severe underweight
Height-for-age	Less than –2 and more than –3 (i.e. 70–79.99% of the norm)	Moderate acute malnutrition (MAM)
Height-for-age	Less than –3 (i.e. less than 70% of the norm) and/ or bilateral pitting edema	Severe acute malnutrition (SAM)

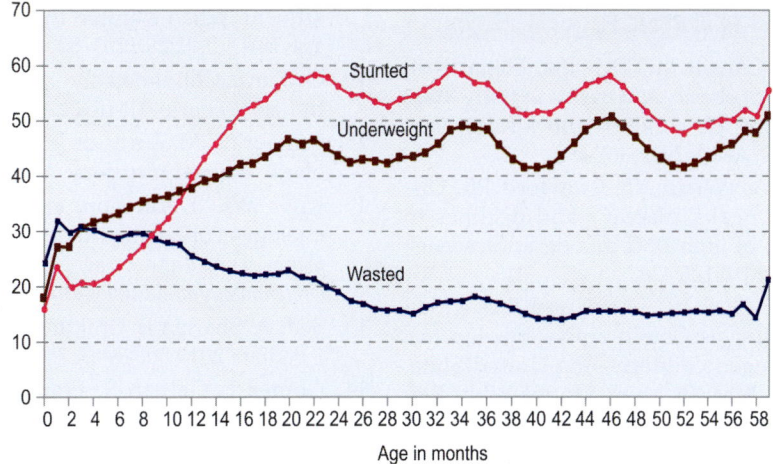

Fig. 4.5: *Showing stunted and underweight children <5 years of age NFHS-3 (2005–2006)*

Malnutrition Among Children

Under Five Years

Almost half of children under age five years (48%) are chronically malnourished. In other words, they are too short for their age or stunted. Stunting is a good long-term indicator of the nutritional status of a population because it does not vary appreciably by the season of data collection or other short-term factors, such as epidemic illnesses, acute food shortages, or shifts in economic conditions.

- Acute malnutrition, as evidenced by wasting, results in a child being too thin for his or her height. One out of every five children in India under age five years is wasted (20%).
- Forty-three percent of children under age five years are underweight for their age.

Underweight status is a composite index of chronic or acute malnutrition. Underweight is often used as a basic indicator of the status of a population's health.

Achievements NFHS-2 to NFHS-3

- The percentage of children who are too short for their age (stunted) decreased by less than one percentage point per year over the seven years between the two surveys, from 51% in NFHS-2 to 45% in NFHS-3.
- The percentage of children who are underweight also decreased, but only by three percentage points. Over this period, the percentage of underweight children decreased by 4 percentage points in urban areas, but by less than 2 percentage points in rural areas.
- Wasting (low weight-for-height) among young children has actually become somewhat worse over time, increasing from 20% in NFHS-2 to 23% in NFHS-3. The increase in wasting is a consequence of the fact that there was a some what greater improvement in stunting than in underweight during this period.

NFHS 2005-06 to Naandi Foundation 2011

- According to the 2011 Hunger and malnutrition survey conducted by the Naandi Foundation, 42% of Indian children under five years old are underweight-almost double the rate of sub-Saharan Africa.
- The survey, which examined the nutritional status of almost 110,000 children across the country, said the consequence of this 'nutrition crisis' were enormous.

REFERENCES

1. Ogden C, Carroll MD, Kit BK, Flegal KM. Prevalence of obesity and trends in Body Mass Index among US children and adolescents, 1999–2010. JAMA 2012;307:483.

2. Madsen KA, Weedn AE, Crawford PB. Disparities in peaks, plateaus and declines in prevalence of high BMI among adolescents. Pediatrics 2010;126:434.

3. Centers for Disease Control and Prevention (CDC). Vital signs: Obesity among low-income, preschool aged childern—the United State. 2008–2011. MMWR Morb Mortal Wkly Rep 2013;62:629.

4. Claive Wang Y, Gortmaker SL, Tareras EM. Trends and racial disparities in severe obesity among the US children and adolescents. 1976–2006. Int J Pediatr Obes 2011;6:12.

5. Cunningham SA, Kramer MR, Narayan KM. Incidence of Childhood Obesity in the United States. N Engl J Med 2014;370:403.

6. Stamatakis E, Wardle J, Cole TJ. Childhood obesity and overweight prevalence trends in England: Evidence for growing socioeconomic disparities. Int J Obes (Lond) 2010;34:417.

7. Khadilkar VV, Khadilkar AV, Cole TJ, Sayyd MG. Cross sectional growth curves for height, weight, and body mass index for affluent Indian children, 2007. Indian Pediatr 2009; 46: 477–85.

8. Agarwal DK, Agarwal KN. Physical growth in affluent Indian children (birth– 6 years). Indian Pediatr 1994;31:377–413.

9. Agarwal KN, Agarwal DK. Upadhyay SK, et al. Physical and sexual growth pattern of affluent Indian children form 5 to 18 years of age IBID 1992;29:1203–82.

10. Gedam S. Childhood obesity-challenges in the Indian Scenario. IJMRR 2013;1:1.

11. Kapoor Y. Majority of Indian children are obese. themedguru.com, August 25, 2010.

12. Klish WL. Definition; epidemiology; and etiology of obesity in children and adolescents. *www.uptodate.com/content/definitio-epidemiology-and-etiology-in chil... 3/13/2014.*

13. Seth A, Sharma R. Childhood obesity. Indian J Pediatr 2013;80:309–317.

14. Gomez F, Galvan R, Frank S. Mortality in second and third degree malnutrition. J Trop, Pediatr 1956;2:77–83.

15. Seoane N and Latham MC. Nutritional anthropometry in the identification of malnutrition in children. *J Trop. Pediatr. Environ. Hlth.,* 1971; 17:98–104.

16. Waterlow JC. Classification and Definition of Protein-Calorie Malnutrition. BMJ 1972;3:566–69.

17. Cole TJ, Flegal KM, Nicholls D, Jackson AA. Body mass index cut-offs to define thinness in children and adolescents: international survey. BMJ 2007;335(7612):194.

18. Natale V, Rajgopalan A. Worldwide variations in human growth and the World health Organization growth standards: A systematic review. BMJ Open 2014;4:1–11.

19. WHO Multicenteric Growth Reference Study Group: WHO child growth standards based on length/height, and weight and age. Acta Pediatr 2006; Suppl 450:76–85.

• Z-score Growth Charts for estimation of Malnutrition <5 years of Age. *See* Chapter 12, pages 110–114.

5

Variations in Pubertal Development and Common Pubertal Problems

Anju Seth

The term *Puberty* refers to the continuum of somatic and sexual changes between childhood and adulthood, leading ultimately to attainment of sexual capacity. The various related terms used commonly in clinical practice include. *Thelarche, which* refers to onset of breast development. *Pubarche* to onset of sexual hair growth. *Menarche* to onset of menstruation and *Spermarche* to appearance of sperms in the seminal fluid.

The term *Gonadarche* refers to onset of pubertal function of gonads that produce sex hormones responsible for pubertal changes, while *adrenarche* is onset of adrenal androgen production. The onset of puberty is quite variable among individuals; the normal age ranges from 8 to 13 years in girls and 9 to 14 years in boys. The developmental stages of puberty follow a well-described sequential pattern, involving concurrent development of gonads, internal and external reproductive organs, appearance of secondary sexual characteristics, growth spurt, changes in body composition and skeletal maturity.

In this chapter, we discuss about physiological variations in pubertal development and issues related to pubertal development occurring too early or too late.

NORMAL VARIATIONS IN PUBERTAL DEVELOPMENT

Unilateral Breast/Testicular Development

Some asymmetry at onset of breast development is common, leading to a difference of one stage advancement in the two sides. This asymmetry among the two sides may sometimes persist into adulthood though the difference may be more obvious at pubertal onset and less perceptible during later stages of development. Very rarely unilateral breast enlargement may be a manifestation of *Poland syndrome*, consisting of unilateral muscular and osseous anomalies of the thorax and upper limb with defects of the ipsilateral breast and nipple or due to a fibroadenoma. In such cases ultrasonography offers a simple, non-invasive tool for diagnosis. There is no role of FNAC or excision biopsy in diagnosing unilateral breast enlargement.

A similar asymmetry may also be noted in testicular enlargement in boys at pubertal onset.

Premature Thelarche

The term premature thelarche refers to isolated breast development in a girl younger than 8 years of age. It may begin unilaterally or bilaterally. Premature the-larche may occur at any age, though two distinct patterns may be observed:

1. Most girls present *during the first two years of life*, when it is usually a persis-tence of, or an increase in breast tissue present at birth. It is caused by per-sistence of infant physiology, when gonadotropin levels are normally elevated, or due to oversensitivity of the breast tissue to small amounts of estrogens secreted secondary to gona-dotropin elevation during this period. Usually, such breast development may regress by two years.

2. The second age period when it may be noted is *after 6 years*, when it is likely to be due to minimal activation of the hypothalamic-pituitary-ovarian axis with predominant FSH secretion. The breast development may regress after a few months or it may persist in pro-portion to somatic growth until normal true puberty begins at a normal age.

In either case, the breast development occurs in isolation. The affected girls do not have a pubertal growth spurt. Their height is thus within the normal range for age and genetic potential. They do not have pubic hair development or menstruation and the bone age is normal. An USG, if performed, does not show any pubertal changes in uterus though sometimes a few small follicular cysts may be seen in the ovaries. Serum estradiol levels are in the normal pre-pubertal range, i.e. <10 pg/ml. Serum FSH levels may be in the high pre-pubertal range, though the response to gonadotropin stimulation test is clearly pre-pubertal.

While evaluating a girl with premature thelarche it must be kept in mind that it may be the first discernible sign of precocious puberty. Thus, these girls should be under careful follow-up to watch for further pubertal progression. A thelarche variant which progresses slowly/intermittently over time along with growth acceleration and skeletal maturation has also been described.

Premature Pubarche

Premature pubarche is characterized by appearance of pubic and/or axillary hair before 8 years of age in girls and 9 years of age in boys. It may be accompanied by adult type body odor, oily skin or mild acne. This development is not accompanied by breast or testicular enlargement, clitoromegaly or an accelerated growth, though affected children may be slightly advanced in height and osseous maturation. Premature pubar-che may be idiopathic, due to apparent sex-ual hair follicle hypersensitivity to the normal traces of androgen of early adrenarche, or due to premature adrenarche. Premature adrenarche ordinarily is diagnosed when premature pubarche is accompanied by a mild elevation of plasma dehydroepi-androsterone sulfate (DHEAS), typically to 40–130 μg/dl which is above the upper limit for normal preadrenarchal children and in the range for normal early pubertal children. However, testosterone and androstenedione levels are not elevated and the gonadotropin response on GnRH stimulation test is pre-pubertal.

Other conditions that can cause pre-mature pubarche include mild errors of steroidogenesis, and testosterone producing adrenal and gonadal tumors. In such con-ditions, other features of systemic androgen excess such as marked growth acceleration, clitoral/phallic enlargement, cystic acne or advanced bone age may be seen. Addi-tionally children diagnosed to have pre-mature adrenarche should be followed up to watch for progressive virilization. Close observation of pubertal development, height velocity, and bone age is also indicated accompanied by lab testing if indicated.

Though thought to be benign, premature pubarche often is associated with obesity and acanthosis nigricans indicative of underlying hyperinsulinemia, and an association between premature pubarche and subsequent risk of polycystic ovary syndrome, hyperandrogenism and the metabolic syndrome has emerged, especially in children born small for gestational age.

Pubertal Gynecomastia

The term gynecomastia refers to benign proliferation of glandular tissue in *males*, leading to enlargement of one or more breasts. Pubertal gynecomastia is a very common condition, and may be seen in as many as 60% adolescents. The usual age of onset is 10–12 years, the peak incidence being at 13–14 years of age, coinciding with SMR stage 3–4 of sexual development. It is a transient condition that resolves spontaneously in a majority of cases over two years or so.

The sub-areolar tissue is felt as a distinct nodule, non-adherent to skin or underlying tissue. In most cases, the nodule is <4 cm in size, resembling SMR stage 2–3 of female breast development. It must be differentiated from lipomastia, i.e. subcutaneous fat collection commonly seen in obese boys, where no distinct nodule can be felt.

Pubertal gynecomastia results from imbalance between stimulatory estrogen action in relation to inhibitory androgen action on the breast. This may be due to increased production of estrogen from the testis/adrenal gland, decreased production of androgens, increased binding of androgens to sex hormone binding globulin as compared to estrogens or an altered androgen metabolism. *In obese boys* there is increased peripheral conversion of androgen into estradiol by aromatase enzyme present in the adipose tissue. This explains the higher incidence of gynecomastia in these boys. Drugs like cimetidine, spironolactone and flutamide displace androgens from androgen receptors thereby increasing the estrogen effect on breast which in turn leads to gynecomatia.

Pubertal gynecomastia must be differentiated from pathological conditions that may lead to gynecomastia in this age group. These include primary or secondary hypogonadism, feminizing testicular/adrenal tumours, liver dysfunction and drugs like INH, phenytoin, ketoconazole and ACE inhibitors in addition to those mentioned above. Prepubertal occurrence of gynecomastia in an adolescent who otherwise shows few/no other signs of puberty or a mid adolescent female degree of gynecomastia is unusual and should prompt evaluation for a underlying pathology. Likewise, presence of genital ambiguity or undescended testis may indicate an underlying disorder of sexual differentiation and should be evaluated accordingly.

Pubertal gynecomastia is usually self-limiting and requires only reassurance and watchful follow-up. In gynecomastia >4 cm or that associated with pain/social embarrassment medical therapy with antiestrogens like clomiphene or tamoxifen may be used. Drug therapy is most effective if given for a gynecomastia of less than 4 years duration. Long standing gynecomastia may have more of fibrotic tissue and may respond poorly to drugs.

Precocious Puberty

'Precocious puberty' refers to early, progressive sexual development, accompanied by rapid linear growth and skeletal maturation ahead of chronological age. In girls, this term is applied if onset of breast development occurs at <8 years, pubic hair at <9 years or menses at <9.5 years of age. Precocity is much less common in boys and is said to be present if secondary sexual characters appear at <9 years of age.

Precocity is broadly of two kinds: *central*, due to early activation of hypothalamo-pituitary-gonadal axis. *Peripheral*, when it occurs under the influence of sex hormones independent of gonadotropin production from the pituitary.

Central Precocious Puberty

In central precocious puberty (CPP), pubertal development is qualitatively and quantitatively like a normal puberty, except for its occurrence at an earlier age. Thus, there is concordant development of gonads, internal organs and secondary sexual characters along with an associated height spurt and advancement in skeletal maturation. Secondary effects like body odour, acne, mood swings may also be present. CPP is a lot more common in girls and is idiopathic in >90% cases. It is less common in boys, in whom one is more likely to identify an underlying organic cause. Hypothalamic hamartoma, a developmental malformation with disordered neuronal migration is the most common organic cause of CPP in both sexes. Other causes include CNS infections, trauma, malformations (hydrocephalus, arachnoid cyst, neural tube defects), irradiation, tumors (optic/hypothalamic glioma, craniopharyngioma, neurofibromatosis) and cerebral palsy. Girls born small-for-date or those with dysmorphic syndromes like Russell Silver are more likely to have an early puberty.

Peripheral Precocious Puberty

Peripheral precocious puberty (PPP) accounts for <1/4th of all children with precocity and occurs in response to abnormal sex hormone production/extraneous administration. The high levels of sex hormones lead to development of target tissues like breasts and external genitalia while gonadotropin production from pituitary is inhibited by high serum sex hormone concentration. Thus, there is no testicular/ovarian development which remain pre-pubertal. Like in CPP, affected children exhibit accelerated growth rate and advanced bone age. PPP is usually isosexual, though occasionally contra-sexual development may be seen depending upon the type of hormones being produced. Disorders commonly associated with PPP include benign ovarian follicular cysts, McCune-Albright syndrome and granulosa cell tumours in girls. In boys, the causes

include congenital adrenal hyperplasia (CAH), familial gonadotropin independent puberty, hCG producing germ cell tumors and masculinizing testicular, adrenal or hepatic tumors. Causes of contra-sexual precocity include congenital adrenal hyperplasia, virilizing ovarian and adrenal tumors in girls and estrogen producing adrenal tumors or exogenous estrogen administration in boys.

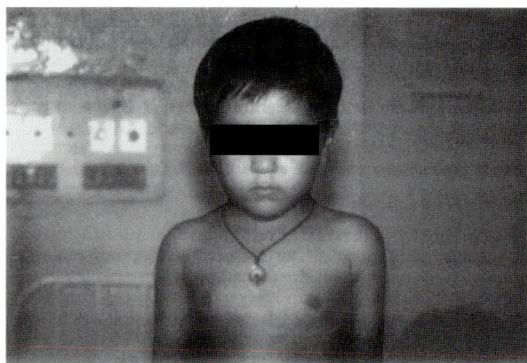

Fig. 5.1: *Child with McCune-Albright syndrome*

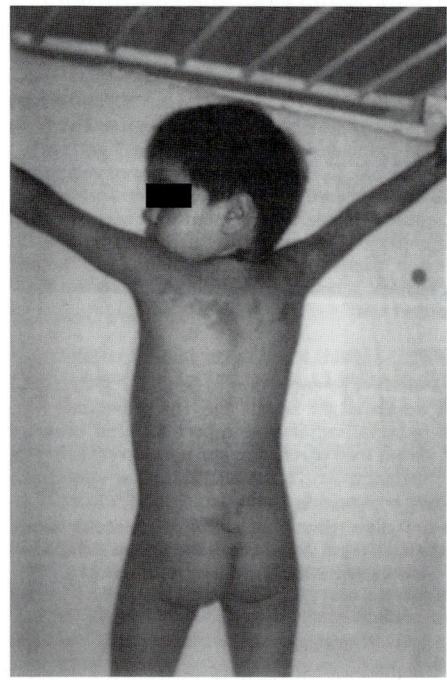

Fig. 5.2: *McCune-Albright syndrome showing café au lait spots*

An important cause of PPP, especially in girls is prolonged untreated hypothyroidism wherein high serum TSH level acts on FSH receptors due to a structural similarity between FSH and TSH, and leads to production of ovarian cysts, which then produce estrogen. Unlike other causes of precocity, girls with hypothyroidism have growth retardation and delayed bone age.

Occasionally, CPP may get triggered in children with PPP when advanced bone age and premature release of sex hormones provokes early maturation of HPG axis.

Evaluation

A detailed history should be elicited in a child presenting with precocious puberty including:

- Time of onset of symptoms, their sequence of appearance and rate of progression
- History of neurological/visual symptoms
- Previous history of diseases/conditions leading to neurological damage
- Developmental history
- Drug exposure
- Family history of precocity, neurofibromatosis
- Physical examination should include:
  - Height and weight in relation to reference standards and genetic potential
  - Description of pubertal stage including testicular volume and asymmetry, and penile length in boys and palpation of breast for galactorrhea in girls
  - Features of contra-sexual development: Hirsutism, clitoromegaly and acne in girls and gynecomastia in boys
  - Neurocutaneous markers, café au lait spots
  - Thyroid enlargement and signs of hypothyroidism
  - Abdominal mass, hepatomegaly
  - Color of vaginal mucosa at the entroitus—glistening and red (non-estrogenized) or dull pink with mucosal secretions (estrogenized)
  - Fundus and neurological examination.

Laboratory Investigations

Investigations are needed to confirm presence of precocious puberty, to differentiate CPP from PPP and to find the underlying etiology, if any. Bone age estimation and an USG to look for pubertal changes in internal organs should be done for all cases. If skeletal age is appropriate for chronological age and ovaries and uterus are pre-pubertal, presenting features may be normal pubertal variants. This is confirmed by documenting prepubertal level of sex hormones, i.e. serum estradiol and testosterone levels <10 pg/ml in girls and <25 ng/dl in boys respectively. These children may be followed up clinically watching for height velocity and progression in pubertal signs.

Indications for further work up are

- Presence of >1 sign of precocity
- Evidence of increased height-for-age and genetic potential
- Advanced bone age >2SD above chronological age
- Progression in pubertal stage during follow-up.

These children require a basal LH, FSH, estradiol/testosterone level. A basal LH of >0.6 IU/L and a LH/FSH ratio >1 indicate pubertal gonadotropin levels. A GnRH stimulation test should be done if basal levels are inconclusive. A post-stimulated LH >6 IU/L, with pubertal sex steroid levels confirms presence of CPP. Normal pubertal variants like isolated thelarche often show a predominant FSH response to GnRH stimulation. In PPP, on the other hand, LH and FSH levels are suppressed while sex steroid levels are high. Thyroid function should be assessed when precocity is associated with growth arrest and a delayed bone age.

All boys with CPP, and girls <6 years age or those with neurological signs irrespective of age must undergo an MRI skull focusing on hypothalamo-pituitary area to look for an underlying cause. Subjects with PPP should undergo an USG abdomen for evaluation of adrenals, ovaries and liver.

In girls with virilization, serum DHEAS level is useful to differentiate premature pubarche from true virilization and 17-OH progesterone levels to exclude congenital adrenal hyperplasia.

Treatment

Untreated precocious puberty leads to early pubertal maturation including commencement of menses and a compromised adult stature. Management involves treatment of underlying cause where possible like hypothyroidism, CAH or a tumor. GnRH agonists are the therapeutic agents of choice in children with CPP exhibiting rapid progression in puberty with significant advancement in bone age and therefore a compromised height potential. Monthly/3 monthly depot injections for intramuscular use are available that need to be administered on a continuous basis till the child reaches an age suitable for pubertal progression. Suppression of gonadotropins to pre-pubertal levels needs to be documented and these patients kept under regular follow-up with periodic evaluation of growth, signs of puberty and bone age maturation. Medroxyprogesterone acetate is another, more economical agent effective in stopping menses and slowing breast development in girls with CPP, though it does not improve the height potential.

Drugs found useful in treatment of PPP due to McCune-Albright syndrome include ketoconazole, spironolactone and aromatase inhibitors. Boys with familial gonadotropin independent puberty respond best to aromatase inhibitors combined with anti-androgen biclutamide. Most ovarian cysts regress spontaneously but would require follow-up. Surgery may be needed for some depending upon size and number of loculations. Psychosocial support to the bewildered child and the anxious family are an essential component of management.

Delayed Puberty

Puberty is considered to the delayed if there is no onset of breast development by 13 years, no pubic hair by 14 years or no menarche by 15 years in girls, or if there is no testicular enlargement by 14 years or pubic hair by 15 years in boys.

Poor pubertal progression is said to be present when >4 years elapse between the first sign of puberty and menarche in girls, or completion of genital growth in boys. The term 'pubertal arrest' is used if there is no progress in puberty over 2 years. In contrast to precocious puberty, delayed puberty is much more common in boys.

Etiology

A delay in onset of puberty is usually temporary, due to a delayed activation of HPG axis. Uncommonly, it may be a manifestation of a disorder causing permanent pubertal failure. Important conditions causing transient delay in puberty include:
- Malnutrition
- *Chronic systemic disorders:* Cardiac, hematological, pulmonary, gastro-intestinal, renal
- *Endocrine disorders:* Hypothyroidism, GH deficiency, glucocorticoid excess, uncontrolled diabetes, hyperprolactinemia
- *Drugs:* Corticosteroids, heroin/opioids
- *Psychiatric disorders:* Anorexia nervosa, psycho-social dwarfism
- Constitutional delay

Delayed puberty is unusual in girls and should prompt a systematic search for an underlying cause.

Permanent pubertal failure could be due to a defect at pituitary/hypothalamus level leading to poor production of gonadotropins (hypogonadotropic hypogonadism), or due to gonadal failure, in which situation gonadotropin levels tend to be high due to lack of feedback inhibition to the pituitary (hypergonadotropic hypogonadism). Important conditions leading to former include congenital panhypopituitarism or isolated gonadotropin deficiency, genetic syndromes such as Laurence-Moon-Beidl, syndrome (LMBS), Prader-Willi (PW) syndrome, Noonan syndrome, or acquired damage to pituitary/hypothalamus secondary to trauma, tumors, inflammation, infiltrative disorders (thalassemia, hemosiderosis), etc. Primary gonadal failure in boys can be due to genetic causes like Klinefelter syndrome and mixed gonadal dysgenesis, due to defects of testosterone biosynthesis, or acquired causes including testicular torsion/trauma. In girls, the most important disorder to be considered is Turner syndrome. Other causes include complete/mixed gonadal dysgenesis, 46 XY complete androgen insensitivity, and autoimmune ovaritis. In either sex infections like mumps and coxsackie, chemoradiotherapy can cause gonadal failure.

Clinical Evaluation

The objective of evaluation is to look for an underlying systemic disorder that could delay puberty and to rule out a disorder that could lead to permanent pubertal failure. Salient points to be elicited in history include:

- Chronic diseases in past
- Review of past height and weight records
- Birth or later head trauma
- Developmental history
- Other hormone deficit/excess
- Visual symptoms, recent growth failure
- Family history of delayed puberty in parents, consanguinity, hypogonadism or infertility, anosmia
- Exposure to radiation/chemotherapy/surgery
- Abnormal eating pattern

Examination should include

- Weight, height, US:LS ratio
- Anemia and other nutritional deficiencies
- *Genitalia:* Males—location and size of testis; Females—clitoromegaly, inguinal hernias/mass in labio-scrotal folds

- *Signs of early puberty:* Breast bud in girls, testis $\geq$2.5 cm or $\geq$4 ml in males. Testes $\leq$1 cm and soft-indicative of hypogonadal state
- Gynecomastia in boys
- Features suggestive of chronic diseases
- *Neurological examination:* Fundus, visual field, sense of smell
- *Stigmata of syndromes:* LMBS, PW, Turner, Klinefelter
- Midline facial defects.

Investigations

The initial lab investigations should rule out a systemic cause. These include: Hemogram, kidney and liver function tests, urine and blood pH, electrolytes, celiac serology, thyroid function and other investigations based on clinical cues. The next step is evaluation of bone age and serum levels of LH, FSH, testosterone/estradiol and prolactin. A delayed bone age suggests chronic systemic disease or constitutional delay. If the bone age is commensurate with onset of puberty or beyond, i.e. >10–11 years for girls and $\geq$12–13 years for boys, presence of elevated gonadotropins in absence of signs of puberty indicates presence of gonadal failure, which has led to a secondary elevation in gonadotropin levels due to lack of feedback inhibition. In these cases karyotype should be obtained. Other investigations that may be needed include testicular biopsy in boys, and USG pelvis and work up for autoimmune disease in girls.

If the bone age is retarded, gonadotropins are low and systemic diseases are ruled out, the main differential is between constitutional delay and permanent hypogonadotropic hypogonadism. This is difficult to establish in a young patient. A GnRH stimulation test in this situation will show a pubertal response in subjects destined to enter puberty in next 6 months or so, while in those with hypogonadotropic states the response is pre-pubertal.

However, a firm differentiation between the two states is often difficult. Therefore, the practical approach is to induce secondary sexual characteristics using low dose of sex steroids for a few months. In addition, to inducing sexual development, the sex steroids would also serve to prime the pituitary to produce gonadotropins and thereby facilitate onset of true puberty. This becomes evident if boys exhibit testicular enlargement and girls have breast development beyond that expected by low dose of steroids. Patients with hypogonadism would require MRI skull and work up for specific mutations as clinically indicated.

the same period. The hormone supplement should then be stopped and patient evaluated clinically as well as by estimation of serum sex steroid level to assess if true puberty has supervened. If not, a repeat course may be considered. Failure to enter spontaneous puberty beyond 1 year of hormone replacement warrants evaluation for hypothalamic/pituitary dysfunction. Subjects found to have permanent pubertal failure regardless of underlying cause would require systematic induction of puberty using incremental hormone supplementation followed by maintenance therapy.

Treatment

In the vast majority of cases where pubertal delay is secondary to a systemic disorder, appropriate management of underlying condition and of associated malnutrition heralds the onset of pubertal development. Adolescents (mostly boys) with constitutional delay need to be counseled regarding the benign and transient nature of the disorder. However, if there is distress over height and immature appearance or if differentiation from a permanent hypogonadotropic state is not firmly established, they can be offered a low dose testosterone therapy. This is given as 4 weekly injection of testosterone enanthate 50 mg intramuscularly for 4–6 months. Girls may be given oral ethinyl eastradiol 5 mg daily for

SUGGESTED READING

1. Bordini B, Rosenfield RL. Normal Pubertal Development: Part I: The Endocrine Basis of Puberty. Pediatrics in Review 2011;32:223–9.

2. Bordini B, Rosenfield RL. Normal Pubertal Development: Part II: Clinical Aspects of Puberty. Pediatr Rev 2011;32:281–92.

3. Carel JC, Lahlou N, Roger M, et al. Precocious Puberty and statural growth. Human Reproduction Update 2004;10(2):135–47.

4. Ferreira L, Silveira G, Latronico AC. Approach to the Patient with Hypogonadotropic Hypogonadism. *J Clim Endocrinol Metab* 2013; 98:1781–8.

5. Kakarla N, Bradshaw KD. Disorders of Pubertal Development: Precocious Puberty. Semin Reprod Med 2003;21(4):339–52. DOI: 10.1055/s-2004-815590.

6

Bone Age: Assessment and Significance

M Vijayakumar, Anju Seth

INTRODUCTION

Bone growth occurs in an orderly and predictable sequence of appearance and development of epiphyseal centers, which can be readily observed on a radiograph. This has led to generation of standards for bone maturation at different ages during childhood and adolescence for both sexes. Comparison of skeletal maturity of an individual with these age and sex related standards forms the bases of bone age (BA) assessment, which refers to the age corresponding to the child's skeletal maturity. Importance of BA lies in the fact that it is a better indicator of biological maturity as compared to the chronological age (CA), and many parameters like height velocity, menarche, accrual of muscle mass and bone mineral density correlate better with BA than CA. Assessment of BA is thus an important tool in evaluation of various types of growth, endocrine and genetic disorders and to assess the impact of treatment in these conditions.

Bone Development

Long bones like radius, ulna and meta-carpals develop from a primary center which will develop into the diaphysis and a secondary center that appears at the end of the bone which forms the epiphysis. Metaphysis is the part of diaphysis adjacent to epiphysis. Growth of the bone ceases and final adult stature is attained when osseous structures of metaphysis and epiphysis fuse.

The process of bone calcification begins as early as 8 weeks of intrauterine period. By 13th week of fetal life, most of the primary centers of long bones are well developed into diaphysis which gets completely ossified at birth. Ossification of distal femoral epiphysis begins by 38 weeks *in utero*, that of proximal tibial epiphysis by 40 weeks of gestation and ossification center of head of femur appears by 1 year of life.

Many factors may influence the normal maturational pattern of bones. These include genetic factors, hormones (thyroxin, growth hormones and sex steroids) and nutrition.

Assessment of Bone Age

Radiographs of entire skeleton may appear a logical tool to assess BA, but are neither practical nor fortunately required. In a newborn, a radiograph of the knee joint showing presence of epiphyseal centers of tibia and femur indicates that the BA corresponds to 40 weeks of gestation. Beyond the neonatal period, for clinical purposes, radiograph of the left hand and wrist is adequate and a commonly employed tool for BA assessment since changes associated with progressive skeletal maturity are reasonably reflected in a radiograph of the hand.

Normal Sequence of Skeletal Development Observed on a Radiograph of Hand (Figs 6.1 and 6.2)

i. First ossification center that appears in the X-ray of hand and wrist is that of

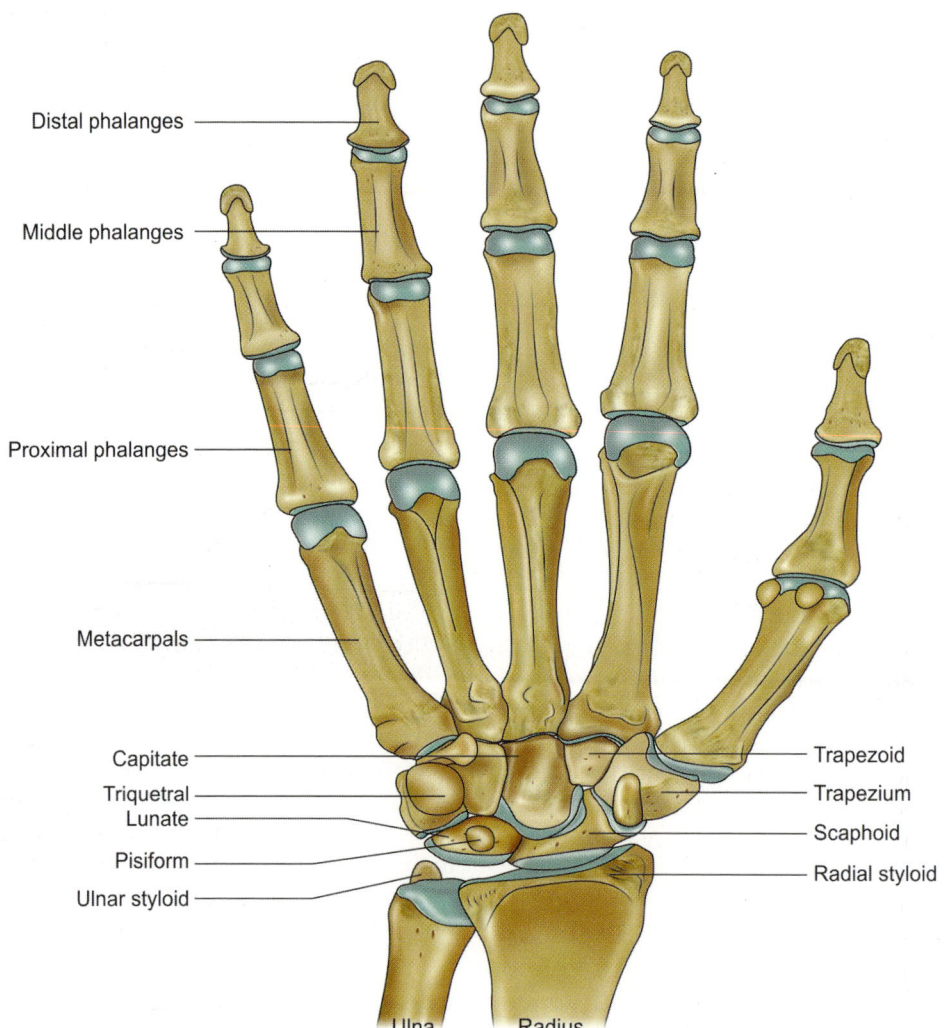

Fig. 6.1: *Bones of the hand and wrist, are sufficient to assess useful bone age in clinical practice*

capitate, closely followed by **hamate** at 3 months of age.

ii. At about 1 year in girls, ossification center in the **distal epiphysis of radius** appears. During 1–3 years age, ossification centers of **metacarpals and phalanges** develop.

iii. The 3rd carpal bone (**triquetral**) develops by 2 years in girls.

iv. Subsequently, ossification center of **lunate** appears at 3 years, and **trapezium** at 4 years.

Boys tend to lag behind girls by 6 months during this period.

v. Subsequently, ossification center for **trapezoid** appears at 5 years in girls, 6 years in boys and **scaphoid** at 6 years in girls, 7 years in boys.

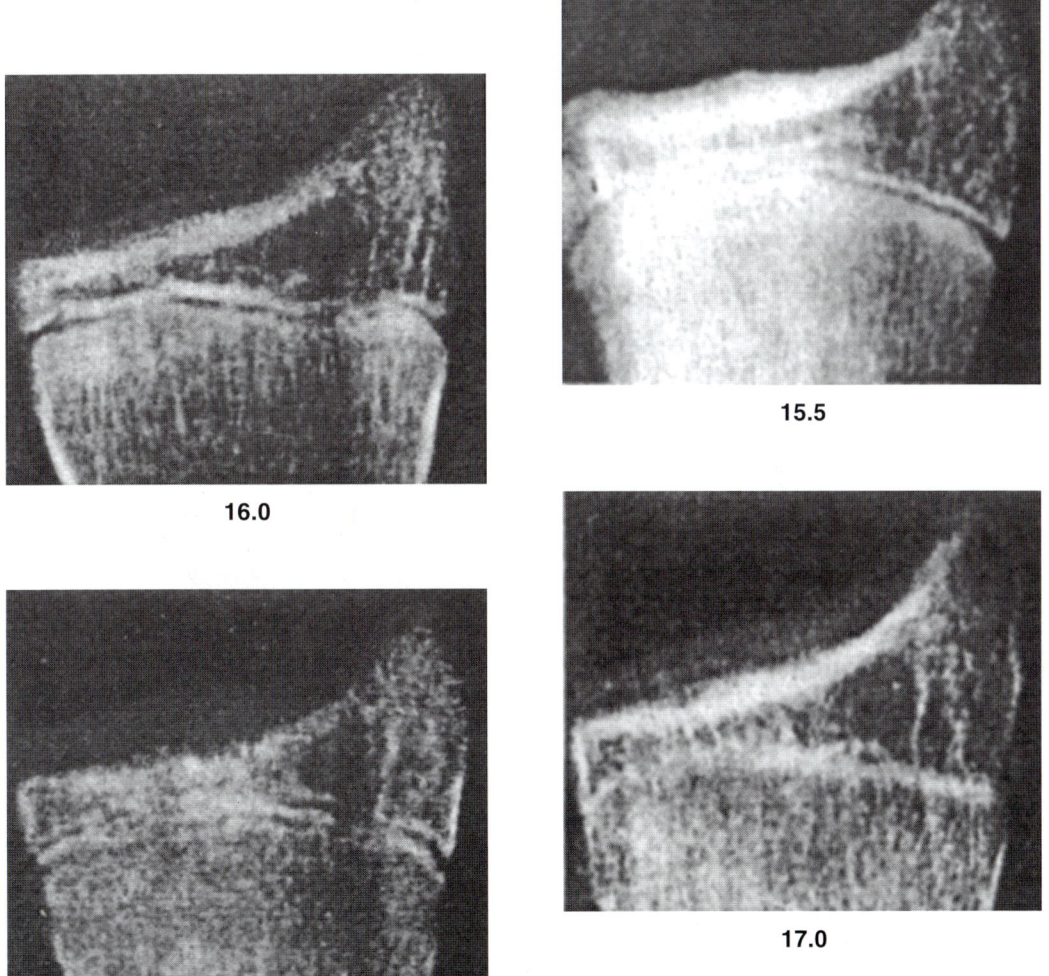

15.5

16.0

17.0

Epiphysis unite

18.0

Fig. 6.2: *The distal end of the left radius in boys aged 15.5–18.0 years, with diaphyseal width kept constant (Greulich–Pyle, 1969)*

vi. **Distal epiphysis of ulna** also appears during this period (5 years in girls, 6 years in boys).

vii. During early puberty, **pisiform** (9 years in girls, 10 years in boys) and **sesamoid** in the tendon of abductor pollicis appear (11 years in girls, 12 years in boys).

viii. Simultaneously, epiphysis of metacarpals and phalanges continue to grow and their width become more than the metaphysis.

ix. As the puberty advances, progressive fusion of epiphysis to metaphysis occurs in the long bones, which begin in the phalanges and metacarpals. Epiphyseal fusion of ulna followed by radius occurs at 15–17 years in girls and 17–19 years in boys.

Technique for Obtaining a Radiograph for Bone Age Assessment

Correct positioning of the hand and wrist while obtaining the radiograph is important since incorrect positioning may give an inaccurate image of the bone's shape. The hand should be placed with palm facing downwards in contact with the cassette. Axis of the forearm should be in direct line with the axis of the middle finger. Upper arm and forearm should be in the same horizontal plane. Fingers should not be touching each other and thumb should be placed at an angle of 30° with the first finger. X-ray tube should be kept half way between the tip of the fingers and distal end of the radius (above the head of third metacarpal). Tube film distance should be 30 inches.

Methods of Assessment of Skeletal Age

BA is assessed by comparing the degree of maturation of various epiphysis in the X-ray of the hand with age related standards. As the age advances, so does the variability in time of appearance and fusion of centers. Thus, mere assessment of appearance and fusion of centers is a crude method of bone age assessment and has limited clinical utility.

In clinical practice, Greulich and Pyle atlas, based on visual evaluation of skeletal development of left hand and wrist, is a commonly used method for assessment of BA. Tanner–Whitehouse (TW3) method is the other frequently used method. In this electronic era, attempts are also being made to develop digital methods to assess the bone age quickly and accurately. In all methods, different reference standards exist for boys and girls.

Greulich and Pyle (GP) atlas

This atlas is divided into 2 separate sets of X-rays, one for boys and the other for girls. Each part contains radiographs of left hand arranged in chronological order. Bone age of a given child is assessed by comparing his/her radiograph with the standards given in the atlas. The method provides BA with intervals of 6–12 months between standards. The X-ray picture that matches the child's X-ray most closely is taken as his/her BA. If the X-ray falls in between two pictures in the atlas, the age is interpreted to be between the ages of the standards it closely resembles.

Tanner and Whitehouse (TW) method

This method is based on individual bone's stages of maturity. Currently the updated version, TW3 is in use. Here, 20 regions of interest (ROI) in different bones are considered for bone age calculation. Each ROI is divided into discrete stages depending on their development and each stage is given a letter (A, B, C, D, E, F, G, H and I). A numerical score is assigned to each stage. By adding the scores of each ROI a maturity score is obtained. This score is compared with the provided charts (separate for boys and girls) and BA is determined.

Though time consuming, TW method gives a more objective assessment of BA as compared to GP atlas where there are more chances of interobserver variation in interpretation. Further, TW3 can differentiate BA up to 1/10th of a year, while GP atlas gives a rough approximation with interval of 6–12 months between standards. TW3 method is thus more sensitive in following small changes in BA.

Computer Assisted Skeletal Age Scores

Recently, various computer assisted techniques for BA assessment have been developed. These include Computer Assisted Skeletal Age Score (CASAS) and Bone Xpert.

Clinical Application of Skeletal Age Assessment

Bone age (BA) assessment provides important clues in diagnosis of various growth disorders. Degree of delay in BA may reflect duration of disease process. BA also helps in predicting adult height and to monitor growth potential over time, especially if a treatment to modify growth/puberty is being given.

BA is delayed in short stature due to chronic systemic diseases, malnutrition, constitutional delay in growth and puberty (CDGP), and various endocrine disorders including growth hormone deficiency and hypothyroidism. It is not generally delayed in skeletal dysplasias since epiphyseal maturation in these disorders remains unaffected. BA is advanced in congenital adrenal hyperplasia, precocious puberty and in overgrowth syndromes like Beckwith-Wiedeman syndrome.

Short Stature

Bone age is an important tool in the diagnosis of short stature. Familial short stature is characterized by BA corresponding to CA or delayed by less than a year. Mild delay (up to 2 years) is commonly seen in malnutrition, systemic diseases and idiopathic short stature. In constitutional delay in growth and puberty, BA is usually delayed by 2–3 years though it correlates well with height age. Endocrine disorders like hypopituitarism and hypothyroidism, on the other hand, are characterized by

marked delay in BA, even more than that expected for height. BA is also useful for assessing response to treatment in these disorders, for example during growth hormone therapy. If increment in height during treatment are larger than increment in bone age, it predicts improved prognosis for final stature.

Pubertal disorders: Exposure to sex hormones accelerates bone maturation. BA is delayed in almost all cases with delayed puberty, while precocious puberty is characterized by marked BA advancement. In normal pubertal variants like premature thelarche and adrenarche, BA advancement is minimal. Serial BA assessment, at least once every year, is recommended during follow-up evaluation of subjects receiving gonadotropin releasing hormone analogue therapy for precocious puberty to assess improvement in predicted height, and to help decide duration of therapy. Slowing of bone maturation on therapy is a sign of adequacy of treatment.

Congenital adrenal hyperplasia (CAH): Adrenal androgens have a marked stimulatory effect on bone maturation. This accelerated bone maturation eventually leads to premature closure of sutures resulting in final short stature. BA assessment is an important component of follow-up evaluation of children with CAH. This helps in titrating the dose of corticosteroids and prediction of final adult height. Inadequate drug dosage results in rapid advancement of BA age due to unsuppressed sex steroid production.

Prediction of adult height: In a normal individual, there is a direct relation between degree of skeletal maturation and time of epiphyseal closure, an event that marks attainment of adult stature. Therefore, there is a significant correlation between BA and the proportion of final stature achieved. More delayed the BA for CA, longer the time before epiphyseal fusion. On the other hand, if the BA is advanced, epiphyseal fusion and attainment of final stature would occur earlier than at the expected CA. Expected adult height can be predicted from a child's height at a particular age and his BA at the time of obtaining the height. Various methods have been developed to assess this, like those by Bayley–Pinneau, Roche, or Tanner–Whitehouse. Prediction of adult height especially in various growth disorders is another important clinical use of assessing BA.

SUGGESTED READING

1. David D Martin, Jan M Wit, Ze'ev Hochberg. The use of bone age in clinical practice, Part 1. Horm Res Paediatr 2011;76:1–9.

2. David D Martin, Jan M Wit, Ze'ev Hochberg. The use of bone age in clinical practice, Part 2. Horm Res Paediatr 2011;76:10–16.

3. Gilsanz V, Ratib O. Hand bone age, a digital atlas of bone age maturity; Springer-Verlag Berlin Heidelberg, 2005.

4. Greulich–Pyle. Radiographic atlas of skeletal development of the hand and wrist (2nd edn). Stanford University Press, Stanford, California.

5. Niemeijera, van Ginneken. Assessing the Skeletal Age from a Hand Radiograph: Automating the Tanner–Whitehouse Method.

6. Tanner JM, Healy JR. Goldstein Assessment of skeletal maturity and prediction of adult height (TW3 method): WB Saunders; 2001.

7

Adult Health Consequences of Being Born with Intrauterine Growth Retardation

PSN Menon

Intrauterine growth retardation (IUGR) and low birth weight (LBW)) are major public health concerns with short-and long-term adverse health consequences. These children have a high risk of developing hypertension and type 2 diabetes mellitus (T2DM) as adults.[1] They are also likely to present with low IQ and short stature.[2] Those with early postnatal catch-up growth run the risk for obesity in childhood and coronary artery disease (CAD), stroke, and T2DM in adulthood.[3, 4] Preterm births are also associated with insulin resistance in children and in adults.[5] Other at-risk groups include those born post-term, twins, offs-prings of gestational diabetic mothers and *in vitro* fertilization. Early detection and optimal management is essential for best possible outcome.

DEFINITION

The World Health Organization (WHO) defines IUGR as those with birth weight below the 10th percentile of the recommended gender-specific birth weight for referenced gestational age. Where gestational age is not available, the birth weight of <2500 g is considered low birth weight and <1500 g as very low birth weight (VLBW). The term IUGR is used if at least two fetal growth assessments are available

and the fetus is not growing appropriately. The term small for gestational age (SGA) is preferred in the absence of evidence of fetal growth.

GROWTH

Most term SGA newborns complete catch-up growth by two years of age.[6] Preterm newborns may take longer to catch-up than term newborns.[7] About 10–15% of those born SGA will continue to have significant short stature (height ≤2SD) during childhood and adult life.[8] Children born SGA hence should be assessed every three months during the first year of life and every six months during the second year. Those who do not show catch-up growth especially during the first six months of life require further evaluation.

SGA children with poor catch-up growth usually have an adequate endogenous growth hormone (GH) secretion in response to pharmacological tests. However, they often have low serum insulin-like growth factor-1 (IGF-1) levels and altered physiological GH secretion patterns.[9]

GH Treatment

The US Food and Drug Administration (FDA) in July 2001 approved GH for the long-term treatment of growth failure in children born SGA who do not have sufficient catch-up growth by age 2 years (height <−2SD).[10] In Europe, GH treatment was approved in June 2003, but with a slightly different criteria—SGA with short stature (height <−2.5 standard deviation score [SDS] and parental adjusted height <−1.0SDS) and failure to show catch-up growth by age 4 years or older.

Dose: The initial GH dose is usually 0.33 mg/kg/week (≈ 47 µg/kg/day or 0.15 IU/kg/day), with dose adjustments based on weight gain. A higher dose is recommended in USA up to 0.48 mg/kg per week (68.5 µg/kg/day). GH therapy increases height velocity in very young SGA children; optimal height is obtained with longer GH treatment before the start of puberty.[11]

Monitoring therapy: Baseline studies including hormonal (thyroid and IGF-1) and metabolic measurements (glucose, insulin, and lipid profile) are mandatory before starting GH treatment. The child should be evaluated every 3–6 months by an experienced pediatric endocrinologist with appropriate dose adjustments.[12] Blood glucose, thyroid function, HbA1c, and IGF-1 should be monitored once a year. Monitoring changes from baseline insulin levels and surrogates of insulin sensitivity are beneficial in the follow-up of these children.

Discontinuation of therapy: Treatment with GH is continued if height velocity increases (>+0.5SDS) in the first year.[13] Treatment should be stopped in adolescence if the height velocity is below 2 cm/year and bone age is >14 years for girls and >16 years for boys (closure of the epiphyseal growth plates).

Adverse effects: Long-term GH therapy is not associated with serious adverse events in short SGA children. They are not more common in SGA than other conditions treated with GH.[13] However, because of increased prevalence of metabolic disturbances and high blood pressure in adults

born SGA, specific attention must be paid to glucose homeostasis and weight gain in short SGA children treated with GH.[1, 3] Discontinuation of long-term GH treatment in SGA adolescents normalized insulin levels (both fasting and stimulated) after a significant increase during GH therapy.[14] It is recommended that IGF-1 concentrations should be monitored and GH dose should be reduced in children with a plasma IGF-1 above +2SD.[15, 16] In a long-term follow-up study of GH therapy in short children born SGA, increased IGF-1 levels were completely reversed after discontinuing GH.[17]

PUBERTY AND GONADAL FUNCTION

The data with regard to initiation, tempo, duration, and progression of puberty in children born SGA are limited. In girls most studies report an age of onset similar to that for children born appropriate for gestation (AGA). However, earlier pubertal onset and menarche as well as late pubertal start, menarche and slow tempo have been reported.[18] Girls with SGA and rapid weight gain in first months of infancy are more likely to have premature adrenarche and menarche than AGA girls.[19] Increased incidence of adrenal and ovarian hyperandrogenism (precocious pubarche) is seen in girls born SGA with early catch-up growth and low insulin sensitivity in a few studies, but not in all.[20]

There is very little information about the long-term effects of IUGR on hypothalamic-pituitary-gonadal function in males.[21]

METABOLIC CONSEQUENCES IN LATER LIFE

Epidemiological studies have shown that children born with IUGR and prematurity are at high risk to develop impairment of carbohydrate metabolism (glucose intolerance, hyperinsulinism and T2DM) and cardiovascular disease (hypertension, dyslipidemia, cerebrovascular disease and coronary heart disease) during adulthood.[22, 23] In addition to the metabolic syndrome characterized by obesity, T2DM, hypertension, stroke and coronary artery disease (CAD), LBW is associated with chronic obstructive airway disease, osteoporosis, and hyperuricemia.

Fetal Programming

The occurrence of the metabolic health consequences in those born SGA has been extensively researched in recent years. Fetal programming has a vital role as the development of various organs during fetal life occurs at critical periods. An environmental insult (e.g. maternal undernutrition or illness) at a critical period can bring about long-term permanent changes in the structure and function of organ. Growth restraint may ensue with undernourished but proportionately small babies born in early gestation or LBW and disproportionate babies in late gestation. These changes have the potential to become maladaptive as the infant grows into adulthood and be the basis of the developmental origin of adult diseases.[24]

Several postulates are offered to explain the developmental programming and its health consequences. The 'thrifty genotype' hypothesis assumes that thrifty genes with enhanced capacity to store fat, selected during fetal undernutrition, increase the risk of obesity and T2DM when exposed to better food availability in later life.[25] The 'fetal salvage' hypothesis further suggests that during a period of critical intrauterine undernutrition, glucose is preferentially distributed to vital organs such as brain and heart to promote fetal survival. This occurs primarily by development of peripheral and/or hepatic insulin resistance.[26] The insulin resistance becomes maladaptive in a nutritionally abundant post-natal environment leading to abnormal metabolic changes. The insulin resistance becomes pathological with the onset of puberty and obesity.

Insulin and Insulin Resistance

Insulin resistance is demonstrated in all age groups born preterm or LBW. Insulin has several metabolic actions in the hepatic as well as peripheral tissues. The hepatic actions include the uptake, assimilation and storage of glucose, amino acids and fatty acids; inhibition of hepatic gluconeogenesis and glycolysis and glycogen storage. Insulin induces glucose uptake by peripheral sensitive tissues mostly skeletal muscle and adipose tissue. Insulin sensitivity signifies the ability of insulin to stimulate glucose uptake in tissues and suppress hepatic glucose release whereas insulin resistance indicates an abnormal decrease in insulin sensitivity. Insulin resistance may be peripheral (skeletal muscle and adipose tissue) or hepatic; or both. It results in compensatory increase in insulin secretion to maintain euglycemia. The peripheral actions of insulin include cell proliferation, sodium retention, sympathetic stimulation, vascular hypertrophy and endothelial function. Persistent hyperinsulinemia may have excess effect on other organs contributing to dyslipidemia, hypertension and endothelial dysfunction. Long-term, insulin resistance significantly increases the risk of T2DM, hypertension, CAD, stroke, metabolic syndrome and cancer.

Rapid Catch-up and Obesity

The transition from relative insulin sensitivity to insulin resistance is associated with rapid postnatal catch-up weight gain, especially a tendency towards central fat deposition.[1] Visceral adiposity is the most important environmental factor that adversely affects insulin resistance. Undernutrition *in utero* increases the risk of obesity in adulthood. Adipocytes programmed during critical periods may have altered function predisposing to later development of visceral adiposity. LBW and preterm infants who have a greater catch-up growth in the first two years after birth are at risk for increased visceral adiposity. Thus, rapid postnatal weight gain amplifies insulin resistance.

Pathogenesis of Fetal Programming

The pathogenetic mechanisms involved in fetal programming include epigenetics, oxidative stress and mitochondrial dysfunction. Epigenetics refers to the non-genomic alteration in DNA and gene expression that can occur due to environmental factors. The Dutch famine study and the Auckland cohort follow-up demonstrate that genome expression can be influenced over several generations by environmental cues resulting in heritable traits. Environmentally induced changes can result in permanent phenotypic changes.[22, 27] DNA methylation and histone modification are examples of epigenetic mechanisms affected by nutrition. Inadequate diet may result in altered DNA methylation which is dependent on dietary glycine, folate and Vitamin B_{12}. Folate or methyl group deficiency in experimental animals during fetal life or early post-natal period produced long-term DNA and specific gene hypomethylation. Recently in the Pune cohort, low maternal Vitamin B_{12} was shown to predict insulin resistance and adiposity in children. These observations may have therapeutic implications.

Indian Studies

Studies from India have confirmed the observations of various phenotypes of LBW and the potential risk from postnatal environment to develop adverse consequences. Up to 30% of Indian children are born LBW either at term or preterm. The Pune cohort described the thin-fat phenotype with less muscle mass and increased subcutaneous fat.[28] Increase in prevalence of T2DM was noted in adults who were born short with a high ponderal index (visceral obesity) in the Mysore cohort.[29] Thinness at birth along

with accelerated weight gain either after the age of two years (Delhi cohort)[30] or during adolescence (Vellore cohort)[31] predicted an increased risk of T2DM. Another rural follow-up study from Vellore showed that maintaining a healthy active lifestyle avoiding visceral adiposity helped to reduce the risk for T2DM.[32] These studies highlight the importance of postnatal factors (food availability and visceral adiposity) in the development of T2DM in India.

Preterm Infants

Most early studies on fetal origin of adult diseases were performed on SGA cohorts; preterm babies appear similar in many ways. The timing of early environmental insults appears to be similar. A recent study has shown that children born premature have higher insulin levels at birth and in early childhood than those delivered at term.[5] The more premature the baby, the higher are the insulin levels. Thus preterm birth (and perhaps early term birth as well) may be a risk factor for the future development of insulin resistance and T2DM. Both IUGR and preterm infants have an early reduction in insulin sensitivity and altered body composition. The Auckland Cohort Study showed that preterm adults around 36 years of age had a marked increase in fat mass and abdominal adiposity, the latter likely reflecting increased visceral adiposity.[27] In addition a gender dimorphism is also noted with males usually more severely affected. More importantly, a similar increase in abdominal adiposity was observed in the term born offspring of parents born preterm, indicating that adverse outcomes associated with preterm birth may extend to the next generation. These observations highlight the need for improving nutrition during pregnancy, prevention of preterm births, early detection of IUGR and preventing early catch-up growth as important preventive strategies.

KEY MESSAGES

1. Up to 30% of Indian children are born LBW either at term/preterm.
2. Both preterm and SGA newborns show a similar metabolic phenotype with later adult disease.
3. Early environmental adversity appears to lead not only to metabolic alterations in glucose metabolism but also rapid post-natal catch-up growth and an increased risk of adult obesity and T2DM.
4. Most children born SGA recover from their weight and height deficiency. However, 10–15% of the children born SGA will continue to have short stature.
5. GH treatment benefits growth potential in short-stature children born SGA.
6. Improving nutrition in girls, pregnant mothers and preterm neonates, measures to prevent LBW and preterm birth, preventing rapid catch-up growth in those born SGA and educating the public about the importance of a healthy lifestyle are essential measures to prevent/minimize the ill effects of fetal programming in adulthood.
7. Children born SGA should be carefully followed by a multidisciplinary group that includes perinatologists, pediatricians, nutritionists, and pediatric endocrinologists, in order to improve growth, glucose homeostasis, and gonadal function.

REFERENCES

1. Mericq V, Ong KK, Bazaes R, et al. Longitudinal changes in insulin sensitivity and secretion from birth to age three years in small- and appropriate-for-gestational-age children. Diabetologia 2005;48:2609–14.

2. Hokken-Koelega A, van Pareren Y, Arends N. Effects of growth hormone treatment on cognitive function and head circumference in children born small for gestational age. *Horm Res* 2005;64(Suppl 3):95–9.

3. Eriksson JG, Forsen T, Tuomilehto J, et al. Effects of size at birth and childhood growth on the insulin resistance syndrome in elderly individuals. Diabetologia 2005;45:342–8.

4. Eriksson JG, Forsen T, Tuomilehto J, et al. Catch-up growth in childhood and death from coronary heart disease: longitudinal study. Br Med J 1999;318:427–31.

5. Wang G, Divall S, Radovick S, et al. Preterm birth and random plasma insulin levels at birth and in early childhood. J Am Med Assoc 2014;311:587–96.

6. Karlberg J, Albertsson-Wikland K. Growth in full-term small-for-gestational-age infants: from birth to final height. Pediatr Res 1995; 38:733–9.

7. Finken MJ, Dekker FW, de Zegher F, Wit JM. Dutch Project on Preterm and Small-for-Gestational-Age-19 Collaborative Study Group 2006. Long-term height gain of prematurely born children with neonatal growth restraint: parallellism with the growth pattern of short children born small-for-gestational-age. Pediatrics 2006;118:640–3.

8. Leger J, Levy–Marchal C, Bloch J, et al. Reduced final height and indications for early development of insulin resistance in a 20-year-old population born with intrauterine growth retardation. Br Med J 1997;315:341–7.

9. de Waal WJ, Hokken-Koelega AC, Stijnen Tet al. Endogenous and stimulated GH secretion, urinary GH excretion, and plasma IGF-I and IGF-II levels in prepubertal children with short stature after intrauterine growth retardation. The Dutch Working Group on Growth Hormone. Clin Endocrinol 1994;41:621–30.

10. Lee PA, Chernausek SD, Hokken-Koelega AC, Czernichow P: International Small-for-Gestational-Age Advisory Board 2003 International Small-for-Gestational-Age Advisory Board consensus development conference statement: Management of short children born small-for-gestational-age, April 24, October 1, 2001. Pediatrics 2003;111:1253–61.

11. Dahlgren J, Albertsson-Wikland K; Swedish Study Group for Growth Hormone Treatment: Final height in short children born small-for-gestational-age treated with growth hormone. Pediatr Res 2005;57:216–22.

12. Wilson TA, Rose SR, Cohen P, et al. Lawson Wilkins Pediatric Endocrinology Society Drug and Therapeutics Committee. Update of guidelines for the use of growth hormone in children: the Lawson Wilkins Pediatric Endocrinology Society Drug and Therapeutics Committee. J Pediatr 2003;143:415–21.

13. Clayton PE, Cianfarani S, Czernichow P, et al. Management of the child born small-for-gestational-age through to adulthood: A consensus statement of the International Societies of Pediatric Endocrinology and the Growth Hormone Research Society. J Clin Endocrinol Metab 2007;92:804–10.

14. van Pareren Y, Mulder P, Houdijk M, et al. Effect of discontinuation of growth hormone treatment on risk factors for cardiovascular disease in adolescents born small-for-gestational-age. J Clin Endocrinol Metab 2009; 88:347–53.

15. Simon D, Leger J, Carel JC: Optimal use of growth hormone therapy for maximizing adult height in children born small-for- gestational-age. Best Pract Res Clin Endocrinol Metab 2008;22:525–37.

16. Chatelain P, Carrascosa A, Bona G, Ferrandez-Longas A, Sippell W. Growth hormone therapy for short children born small-for-gestational-age. Horm Res 2007;68:300–9.

17. Bannink EM, van Doorn J, Mulder PG, Hokken-Koelega AC. Free/dissociable insulin-like growth factor (IGF)-I, not total IGF-I, correlates with growth response during growth hormone treatment in children born small-for-gestational-age. J Clin Endocrinol Metab 2007;92:2992–3000.

18. Ibáñez L, Ferrer A, Marcos MV, et al. Early puberty: Rapid progression and reduced final height in girls with low birth weight. Pediatrics 2000;106:E72.

19. Ibáñez L, Jimenez R, de Zegher F. Early puberty-menarche after precocious pubarche: Relation to prenatal growth. Pediatrics 2006; 117:117–21.

20. Ibáñez L, de Zegher F. Puberty after prenatal growth restraint. Horm Res 2006;65(Suppl 3):112–5.

21. Main KM, Jensen RB, Asklund C, et al. Low birth weight and male reproductive function. Horm Res 2006;65(Suppl 3):116–22.

22. Barker, DJ, Osmond C, Golding J, et al. Growth in utero, blood pressure in childhood and adult life, and mortality from cardiovascular disease. Br Med J 1989;298:564–7.

23. Dunger DB, Ong KK. Babies born small-for-gestational-age: Insulin sensitivity and growth hormone treatment. Horm Res 2005;64 (Suppl 3):58–65.

24. Barker DJ. The fetal and infant origins of adult disease. Br Med J 1990;301:1111.

25. Hales CN, Barker DJP. Type 2 (non-insulin-dependent) diabetes mellitus: The thrifty phenotype hypothesis. Diabetologia 1992; 35:595–601.

26. Hofman PL, Cutfield WS, Robinson EM, et al. Insulin resistance in short children with intra-uterine growth retardation. J Clin Endocrinol Metab 1997;82:402–6.

27. Mathai S, Derraik JG, Cutfield WS, et al. Increased adiposity in adults born preterm and their children. PLoS One 2013 Nov 20; 8(11): e81840. doi: 10.1371/journal.pone.0081840. eCollection 2013.

28. Yajnik CS, Fall CH, Coyaji KJ, et al. Neonatal anthropometry: The thin–fat Indian baby. The Pune Maternal Nutrition Study. Int J Obes Relat Metab Disord 2003;27:173–80.

29. Fall CH, Stein CE, Kumaran K, et al. Size at birth, maternal weight, and type 2 diabetes in South India. Diabet Med 1998;15:220–7.

30. Bhargava SK, Sachdev HS, Fall CH, et al. Relation of serial changes in childhood body-mass index to impaired glucose tolerance in young adulthood. N Engl J Med 2004; 350:865–75.

31. Raghupathy P, Antonisamy B, Geethanjali FS, et al. Glucose tolerance, insulin resistance and insulin secretion in young south Indian adults: Relationships to parental size, neonatal size and childhood body mass index. Diabet Res Clin Pract 2010;87:283–92.

32. Nihal T, Louise G, Pernille P, et al. Born with low birth weight in rural Southern India: what are the metabolic consequences 20 years later? Eur J Endocrinol 2012;166:1–10.

8

Brain Development and Neurodevelopmental Processes: The Role of Imaging Techniques

Sonika Agarwal

Human brain development constitutes striking biologic and functional development of the brain's fiber tracts as well remodeling of cortical and subcortical structures. It is a protracted process, beginning in the third week of gestation and continuing into early adulthood (Stiles, 2008).[1] Recent studies using magnetic resonance imaging (MRI) confirm biological development of brain throughout childhood. Studies also provide evidence that various neural structures and systems exhibit different trajectories of maturation. During the years of infancy and school-age, ongoing maturation of brain's connecting fiber tracts and MRI evidence of these changes correlates with neurodevelopment parameters and behavior (Lebel et al 2008; Westlye et al 2010).[2, 3]

Development is dependent of changes on brain structure and function. Differences in rate and extent of brain maturation are likely to have an effect on development and behavior even within the normal range. The progress of myelination appears to follow a distinctive temporal and spatial pattern (Paus et al 2001).[4] Beginning at birth, myelination starts at the base of the brain with the pons and the cerebellar peduncles and then progresses to the posterior optic radiation and the splenium of the corpus callosum (1–3 months). Around 6 months of age, the myelination continues to move forward to the anterior limb of the internal capsule and the genu of the corpus callosum. Around 8–12 months age, frontal, parietal and occipital lobes begin with the myelination process. Cognitive abilities are thus attained at varying ages across the population, based on the speed with which an infant brain develops and the myelination proceeds.

Before the advent of MRI, it was thought that the biologic development of the human brain was complete by 6 years of age. Recent years have seen the use of MRI and functional MRI to characterize the impact of brain growth on neurodevelopment. MRI of the brain can delineate the proliferation of oligodendrocytes and myelination of various structures based on signal changes, thus facilitate tracking of brain growth in relation to neurodevelopment (Barkovich, 2005).[5] The appearance of brain on MRI changes considerably due to an orderly pattern of myelination in white matter regions, over the first 2–3 years of life. Brain morphological changes past the third year of life are more subtle and were first noted in 1990s after the quantitative morphometric techniques were first applied to human brain imaging. Estimates of gray matter volumes, both in cerebral cortex and subcortical nuclei, were larger in school-age children than in adults (Jernigan et al 1991, Jernigan and Tallal, 1990).[6, 7] These studies indicate that dynamic biologic changes in the brain tissue and structure continue through the early years into adolescence [see Chapter 2 The Adolescent Brain, pages 40 and 41].

Recent morphometric studies by MRI provided insight into more anatomical detail, ongoing myelination and are also employing mapping methods for tracking age related changes in brain growth (Otsby et al 2009; Shaw et al 2008).[8, 9] All these studies established a protracted course of white matter growth and changes in the volume of gray matter in cerebral cortex and deep brain nuclei. Further studies in child and adolescent brain have shown the changes in cerebral cortex, apparent thinning in some areas with selective thickening in other areas of the cerebral cortex (Gogtay et al 2004).[10] The cortical thinning is first evident in the primary sensory-motor cortex, and then progresses to secondary. In the later years of childhood and adolescence it is seen throughout the multimodal cortical areas. Increasing myelination is also speculated to be responsible for some of the cortical thinning in these areas during the years of brain growth through childhood and adolescence (Sowell et al 2002).[11] In summary, MR morphometric studies reveal a complex pattern of brain growth and provided more information about ongoing maturation of fiber tracts.

Diffusion MRI measures the diffusion of water molecules through the tissue. Diffusion tensor imaging (DTI) provides useful measures of magnitude and direction of the diffusion. The physical constant characterizing the water motion is called 'apparent diffusion coefficient' (ADC). Diffusion-based MR imaging techniques provide one of the earliest indicators of tissue injury, showing changes many hours to days before changes are detectable by other imaging methods. These have been increasingly used to detect microstructural state of white matter and developing cortex, and understanding brain injury and neurodevelopment disabilities in the preterm and postnatal brain (Mathur et al 2010, Mukherjee et al 2006).[12, 13] Abnormal diffusion imaging parameters at term equivalent postmenstrual age have been associated with poor neurodevelopmental outcome. In the presence of a normal conventional MRI, an isolated increase in ADC values in central white matter in preterm infants at term equivalent age correlated with a lower developmental quotient at two years corrected age.[14] The central white matter in preterm infants is thought to be uniquely vulnerable to injury and the degree of gestational immaturity if related to this impact, based on imaging studies using DTI and diffusion MR tractography.

Also, conventional MR images can be used for measuring tissue volumes, the technique of volumetric measurement through MR imaging. Overall, prematurity and white matter injury are associated with reduced brain tissue volume and increased cerebrospinal fluid (CSF) volumes. Various studies have demonstrated association of reduction in cerebral cortical gray matter volume, increased CSF volume, decreased total myelinated white matter volume, immature gyral development in premature infants with white matter injury, adding to the research on impact of prematurity on developing brain.[12, 15] Brain metrics and surface-based analysis of cerebral cortex are increasingly being used in research studies to quantify regional abnormalities in the developing brain in relation to neurodevelopmental issues.

In the recent years, advent of functional MRI (fMRI) has offered novel research on association of injury seen with the neurodevelopmental population in the high-risk pediatric population. It can be used to detect neural activation through its sensitivity to local oxyhemoglobin and deoxyhemoglobin levels, thus changing the signal intensity.

These studies are in experimental stage and in future may provide key to understanding variability in compensation of the function and development in the infant population in relation to perinatal and postnatal events and brain injury.

The above imaging techniques are also being used to study behavior and cognitive development in addition to other neurodevelopmental issues and are discussed in Chapter 9.

REFERENCES

1. Stiles J. The fundamentals of brain development: Integrating nature and nurture. Brentwood, CA, USA, Harvard University Press, 2008.

2. Lebel C, Walker L, Leeman A, et al. Microstructural maturation of the human brain from childhood to adulthood. Neuroimage 2008; 40:1044–55.

3. Westlye LT, Walhovd KB, dale AM, et al. Lifespan changes of the human brain white matter: Diffusion tensor imaging and volumetry. Cerebral cortex 2010;20:2055–68.

4. Paus T, Collins DL, Evans AC, et al. Maturation of white matter in human brain: A review of magnetic resonance studies. Brain Research Bulletin 2001;54(3):255–6.

5. Barkovich AJ. Magnetic resonance techniques in the assessment of myelin and myelination. Journal of Inherited Metabolic disease 2005; 28:311–43.

6. Jernigan TL, Traumer DA, Hesselink JR, et al. Maturation of human cerebrum observed *in vivo* during adolescence. Brain 1991;114: 2037–49.

7. Jernigan TL, Tallal P. Late childhood changes in brain morphology observable with MRI. Developmental Medicine and Child Neurology 1990;32:379–85.

8. Otsby Y, Tammes CK, Fjell AM, et al. Heterogeneity in subcortical brain development: A structural magnetic resonance imaging study of brain maturation from 8 to 20 years. The Journal of Neuroscience 2009;29(11):772–82.

9. Shaw P, Kabani N, Lerch JP, et al. Neurodevelopmental trajectories of human cerebral cortex. The Journal of Neuroscience 2008; 28:3586–94.

10. Gogtay N, Giedd JN, Lusk L, et al. Dynamic mapping of human cortical development during childhood through early adulthood. Proceedings of the National Academy of Sciences of the United States of America 2004; 101:8174–9.

11. Sowell ER, Trauner DA, Gamst A, et al. Development of cortical and subcortical brain structures in childhood and adolescence. At structural MRI study. Developmental medicine and Child Neurology 2002;44:4–16.

12. Mathur AM, Neil JJ, Inder TE. Understanding brain injury and neurodevelopmental disabilities in the preterm infant: The evolving role of advanced MRI. Seminars in Perinatology 2010;32(1):57–66.

13. Mukherjee P, McKinstry RC. Diffusion tensor imaging and tractography of human brain development. Neuroimaging Clinics N America 2006;16(1):19–43.

14. Krishnan ML, Dyet LE, Boardman JP, et al. Relationship between white matter apparent diffusion coefficients in preterm infants at term-equivalent age and developmental outcome at 2 years. Pediatrics 2007;120(3):604–9.

15. Inder Te, Warfield SK, Wang H, et al. Abnormal cerebral structure is present at term in premature infants. Pediatrics 2005;115(2):286–94.

9

Plasticity in Developing Brain

Mary Cole

INTRODUCTION

Adolescence is a neuropathic critical period when a fundamental reorganization of brain development takes place. In the last 10–15 years the increasing use of functional MRI and other neuroimaging developments have led to an explosion in the ability to image brain processes as they occur (Blakemore).[1]

This critical period includes intensive myelination, reduction in gray matter volume[2] and synapse pruning.[3] The cerebral cortex reaches its maximal volume soon after birth but maturation of gray matter takes place in the sensory motor cortex before the prefrontal cortex, which serves higher functions including fear response and risk assessment. We know that prefrontal dendritic spine density increases throughout childhood to 2 or 3 times than that seen in adulthood and then decreases after puberty.[4, 5] Consistent MRI findings suggest an increase in white matter volume through childhood and adolescence with adolescents showing a higher percentage volume of white to gray matter in the prefrontal cortex aged 14 compared to age 9. Geidd et al[6] conducted scans every 2 years on 145 healthy subjects in age from 4 to 22 years this showed increased volume of gray matter in the frontal lobe in late childhood with a decline in adolescence, with a peak in temporal lobe gray matter at around 17 years old. Other studies also confirm the decrease in gray matter up to the age of 30 years. The higher percentage volume of the white matter therefore represents both an absolute increase in volume and a proportionate one.

The dorsal medial prefrontal cortex is more highly activated in adolescents and adults.[7] When looking at more complex 'higher' emotions such as guilt and this activation leads to the acquiring of effective mentalization (understanding the emotions of others) in adolescents. In other words, survival emotions (such as fear) are acquired earlier in life but more sophisticated mentalization of other people's emotions is acquired later in humans. However, Moor et al[8] looked at brain activity in adolescents age 10–12, 14–16 and young adults age 19–23 years. They recorded their brain activity while these people were making judgements about the emotions of others based only on photographs of their eyes. In this task, only early adolescents (10–12 years) showed additional involvement and dorsomedial prefrontal cortex over the ventral medial prefrontal cortex.

We know that during puberty volume of the gray matter in hippocampus decreases and the volume of the gray matter in amygdala increases. In women there is a proportionate increase in gray matter in the limbic area, related to estrogen levels and in men there is a negative relationship between testosterone levels and the gray matter in the parietal cortex.[9]

Deprived Children
• Children who had been in an orphanage at any time in their lives had much smaller gray matter volume in the cortex of the brain and had smaller white matter volume than those who had never been in an orphanage. Even if children were placed in loving foster homes, the formerly institutionalized children's gray matter didn't catch up. • White matter, however, seemed to be more resilient. As orphaned children placed in high-quality foster care had the same white matter volume as those who were never in an orphanage.

The next few paragraphs will look at the most active areas of plasticity during adolescence and some of the pathological processes that can occur which affect brain development into adult life.

Medial Prefrontal Cortex

The ventral part of the medial prefrontal cortex is one of the last brain regions to mature during development.

The function of the medial prefrontal cortex is strongly implicated in the etiology of anxiety and mood disorders. Early life stresses (including emotional abuse) is known through both epidemiological and clinical studies to increase the risk of these disorders, possibly mediated through epigenetics and raised glucocorticoid levels.[10] The medial prefrontal cortex is one of the last brain regions to mature[11] and the ventral medial prefrontal cortex regulates responses to fear and anxiety through its connections with the amygdala and peri-aqueductal gray matter. Studies now suggest that there is a link between depressive and anxiety behavior with a specific change in the medial prefrontal cortex as well as changes in the GABA and glutamate systems.[12, 13] Histopathological examination of the brains of patients with mood and anxiety disorders showed reduced gray matter volume, reduced synapses and changes in metabolic activity.[14] These changes can be linked back to early life stress through the effect of stress on the developmental trajectory of the medial prefrontal cortex.[15, 16] This is seen initially during maternal separation and the effect this has on the surges of dopamine expression in the adolescent period. Adolescents with a history of early life stress have also been found to have a reduced volume of the ventral part of the medial prefrontal cortex.[17]

Hippocampus

The hippocampus reaches about 85% of it adult volume before the age of 5 years.[18] For many years it has been known that in lower mammals the environment in which they are reared will induce experience dependant neuroplasticity. Complex environmental and social stimulation (mazes and interreactions) increase dendrite branching and glial cell numbers. Kemperman[19] found that these kind of enriched environments produced statistically significant increase of up to 15% in the granule cell layer of the hippocampus. Conversely, it is suggested that childhood physical and sexual abuse may be associated with diminished hippocampal development.[20] Studies of the hippocampus show that myelination occurs more rapidly in girls than in boys between the ages of 6 and 19.[21] This may lead to different sensitive periods for physical and sexual abuse, leading to different long-term consequences depending on gender.

Volumetric MRI scans from adult women with repeated episodes of childhood sexual abuse have shown that early onset and longer duration of abuse correlates with greater changes in hippocampal volume. Reduced volume of the hippocampus is most strongly related to abuse between the ages of 3 and 5 years old but secondly to abuse between the ages of 11 and 13 years, with the frontal cortex being affected between the ages of 14 and 16 years. This latter change is thought to be mediated through the effects of cortical releasing hormone on cell survival and dendritic branching.[22]

Adolescence is a time of experience dependent adaptive plasticity and this makes experience dependent learning important in shaping cognitive pathways and driving future behavior in a cycle that can become self-perpetuating. We know that up and down regulation of dopamine and glutamate receptors occur in response to psychostimulants such as cocaine. It is now also shown that neuronal circuits themselves, the dendrites and synapses can be modified by exposure to drugs of abuse.[23] Chronic exposure to alcohol also produces dendritic spine changes in the hippocampus[24] this can be delineated in adolescent subjects with alcohol misuse disorders with a reduced hippocampal volume compared to age matched controls.[25] A paper by Carpenter-Hyland et al[26] looks at adaptive plasticity of NMDA (N-Methyl-D-Aspartate) receptors and dendritic spines in the adolescent brain. This confirms that on a cellular level that homeostatic adjustments made in response to the regular use of alcohol in adolescents leads to changes in dendritic spines architecture as well as increased localization of NMDA receptors at the synapse. If these changes become established during a period and adolescent brain plasticity, the addictive behavior can become much harder to treat during adulthood when plasticity is less. This fits in with clinical findings that early alcohol use is more likely to lead to problematic use in adulthood which can be treatment resistant (Chapter 2; page 40).

Corpus Callosum

The nerve fiber connections of corpus callosum are fully developed before birth but myelination and pruning continues into young adulthood. The rostral to caudal myelination pattern suggests that different regions of the corpus callosum might be vulnerable to insults at different time periods. We know the corpus callosum size is reduced in male primates who are isolated in early development compared to socially reared primates.[27] In another example, animals studies show that early beneficial handling[28] that 'handled' male rats had increased corpus callosum volume compared to less stimulated controls. Historically most laboratories have standardized experimental animals of one gender. Only more recently have animal studies began to look at the effects of environmental events across puberty in both sexes. In humans there are naturalistic studies that suggest different vulnerabilities in males and females.

A study by Tiecher et al[29] shows that neglect in boys (mean age 12.9 ± 2.9 years) was associated with a marked decrease in the size of the mid-body and splenium of corpus callosum but with a much smaller effect size in girls. However, in girls, sexual abuse affected the mid-body and isthmus volume, which in sexually abused boys seemed relatively protected. This may be due to gender dependent differences in the time-course of myelination between the sexes and different environmental susceptibility of the sexes to different stresses.

Conclusion

This adaptive plasticity is one of the great therapeutic tools available to the adolescent psychiatrist. Adolescents have a period of developing strength and resilience alongside their potential vulnerability. There may be a set of biologically-based neural changes, especially ones moderating emotions, which drive what may be a normal adolescent tendency towards recklessness, out of kilter with their cognitive function or indeed how the same person would have reacted to a similar risk when a younger child. There is also an anthropological perspective, while puberty is occurring earlier in many industrial societies, the taking on of adult roles is delayed. The task of society and health professionals is to constantly strive to produce an enriching rather than a toxic environment for the developing adolescent brain.

REFERENCES

1. Blakemore S. Imaging brain development. The adolescent brain. Neuro Image 2012;61: 397–406.

2. Banser M, Duman RS. Glial loss in the prefrontal cortex is sufficient to induce depressive-like behaviours. Biol Psychiatry 2008; 64:863–70.

3. Baudin A, Blot K, Verney C, et al. Maternal deprivation induces deficits in temporal memory and cognitive flexibility and exaggerates synaptic plasticity in the rat medial prefrontal cortex. Neuro Biol Learn Mem 2012; 98:207–14.

4. Petanjek Z, Judas M, Simic G, et al. Extrodinary neoteny of synaptic spines in the human prefrontal cortex. Proc Natl Acad Sci. USA 2011;32:13281–6.

5. Sowell ER, Thompsom PM, Holmes CJ, et al. Localising age-related changes in brain structure between childhood and adolescence using statistical parametric mapping. Neuro Image 1999;6(1):587–97.

6. Giedd JN, Blumenthal J, Jeffries NO, et al. Brain development during childhood and adolescence: A longitudinal MRI study. Nat Neurosci 1999;2(10):861–3.

7. Burnett S, Blakemore SJ. Development during adolescence of the neural processing of social emotion. J Cogn Neurosci 2009;29:1294–1301.

8. Gunther Moore B, Op de Macks ZA, Guroglu B, et al. Neurodevelopmental changes of reading the mind in the eyes. Soc Cog Affect Neurosci 2012;7(1):44.

9. Neufang S, Specht K, Hausmann M, et al. Sex differences and the impact of steroid hormones on the developing human brain. Cereb. Cortex 2009;19:464–73.

10. Van Harmlen AL, Van Tol MJ, Van der Wee NJ, et al. Reduced medial prefrontal cortex volume in adults reporting childhood emotional maltreatment. Biol Psychiatry 2010; 68:832–8.

11. Brenhouse HC. Developmental trajectories during adolescence in males and females: a cross-species understanding of underlying brain changes. Neurosci Biobehav Rev 2011; 35:1687–703.

12. Chocyk A, Dudys D, Przyborowska A, et al. Early life stress affects the structural and functional plasticity of the medial prefronatal cortex in adolescent rats. Eur J Neurosci 2013; 38:2089–107.

13. Pascaul R, Zamora-Leon SP. Effects of neonatal maternal deprivation and post-weaning environmental complexity on dendritic morphology of prefrontal pyramidal neurons in the rat. Acta Neurobiol Exp (Wars) 2007; 67:471–9.

14. Dreuets WC, Price JL, Furey MC. Brain structural and functional abnormalities in mood disorders. Implications for neurocircuitry models of depression. Brain Struct Funct 2008; 93:118.

15. Callaghan BL, Richardson R. Early-life Stress affects extinction during critical periods of development: An analysis of the effects of maternal separation on extinction in adolescent rats. Stress 2012;15:671–79.

16. Brenhouse HC. Early life adversity alters the developmental profiles of addiction related prefrontal cortex circuitry. Brain Sci 2013;3: 143–58.

17. Rinne-Albers MA, van der Wee NJ, Lamers-Winklemann F, et al. Neuroimaging in children, adolescents and young adults with pathological trauma. Eur. Child. Adolesc Psych 2013;22:745–55.

18. Geidd JN, Vaituzis AL, Hamburger SJ. Quantitative MRI of temporal lobe of amygdala and hippocampus in normal human development aged 4–18 years. Journal Comp Neurol 1996; 336:223–30.

19. Kempermann G, Kuhn HG, Gage F. More hippocampal neurons in adult mice living in an enriched environment. (Let) Nature 1996; 386:493–5.

20. Stein MB. Hippocampal volume in women victimized by childhood sexual abuse. Psychol Med 1997;27:951–9.

21. Benes FM, Turtle M, Kahan Y, et al. Myelination of key relay zone in hippocampal formation occurs in the human brain during childhood, adolescence and adulthood. Archives of general psychiatry 1994;51:477–4.

22. Brunson KL, Eghbal-Ahmadi M, Bender R et al. Long-term progressive hippocampal cell loss and dysfunction induced through early life administration of corticotrophin releasing hormone reproducing the effect of early life stress. Proc Natl Acad Sci USA 2001; 98:8856–61.

23. Robinson TE, Kolb B. Structural plasticity associated with exposure to drugs of abuse. Neuropharmacology 2004;47:33–46.

24. Tarelo-Acuna L, Olvera-Lortes E, Gonzalez-Burgos I. Prenatal and postnatal exposure to ethanol induced changws in the shape of dendritic spines from hippocampal CA1 pyramidal neurons of the rat. Neurosci Let 2000;286:13–6.

25. De Bellenis, Clark DB, Beers SR, et al. Hippocampal volume in adolescent onset alcohol use disorders. Am J Psychiatry 2000;157:737–44.

26. Carpenter-Hyland EP, Judson-chandler L. Adaptive plasticity of NMDA receptors and dendritic spines: Implications for enhanced vulnerability of adolescent brain to alcohol addiction. Pharmacol Biochem Behav 2007; 86:200–8.

27. Sanchez MB, Hearn EF, Do-D, Rilling JK, et al. Differential rearing affects corpus callosum size and cognitive function of rhesus monkeys. Brain Res 1998;81:38–49.

28. Liv D, Caldji C, Sharma S, Plotsky PM, et al. Influence of neonatal rearing conditions on stress induced adenocorticotropin responses in hypothalamic paraventricular nucleus. J Neuroendocrinology 2000;12:5–12

29. Teicher MH, Dumont NL, Ito Y, et al. Childhood negelect is associated with reduced corpus callosum area. Biol Psychiatry 2004; 56:80–5.

10

Dental Growth

KN Agarwal, Raj Kumar

Deciduous teeth, otherwise known as **milk teeth**, **temporary teeth** and now more commonly **primary teeth**, are the first set of teeth in the growth development of humans. They appear during the embryonic stage of pregnancy at 6th week of tooth development. This process starts at the midline and then spreads back into the posterior region. By the time the embryo is eight weeks old, there are ten buds on the upper and lower arches that will eventually become the primary (deciduous) dentition, development and erupt in the mouth—during infancy. They are usually lost and replaced by permanent teeth, but in the absence of permanent replacements, they can remain functional for many years. In the primary dentition there are a total of twenty teeth: Five per quadrant and ten per arch. The eruption of these teeth begins at the age of six months and continues until twenty-five to thirty-three months of age during the primary dentition period. Usually, the first teeth seen in the mouth are the mandibular central and the last are the maxillary (Table 10.1). The teething age of primary teeth is summarized below:

The process of shedding primary teeth and their replacement by permanent teeth is called exfoliation. This may last from age 6 to 12 years. By age twelve there usually are only permanent teeth remaining.

Functionally 'Primary teeth' are considered essential in the development of the

Teething age of primary teeth		
Primary teeth	Calcification begins	Appearance
Incisors	4 mo (fetal life)	Central 6–12 mo
		Lateral 9–16 mo
Canines	5 mo (fetal life)	16–23 mo
1st molars	6 mo (fetal life)	13–19 mo
2nd molars	6 mo (fetal life)	22–33 mo

Table 10.1. *Pattern of deciduous teeth eruption in Delhi children*

		Age at earliest appearance (months)	Mean (± SD) age at eruption (months)
a)	**Central incisors**		
	Lower	5	8.2 ± 1.8
	Upper	7	9.3 ± 1.9
b)	**Lateral incisors**		
	Upper	7	11.6 ± 1.9
	Lower	8	12.3 ± 2.5
c)	**First molar**		
	Upper	15	16.0 ± 0.9
	Lower	16	16.3 ± 0.5
d)	**Canine**		
	Lower	17	20.4 ± 1.7
	Upper	18	19.9 ± 2.7
e)	**Second molar**		
	Lower	24	27.0 ± 1.8
	Upper	26	28.8 ± 0.9

Agarwal et al 2001

oral cavity. The permanent teeth replacements develop from the same tooth germs as the primary teeth, which provide guides for permanent teeth eruptions. Also the jaw muscles and the formation of the jaw bones depend on the primary teeth to maintain proper spacing for permanent teeth. The roots of primary teeth provide an opening for the permanent teeth to erupt. The primary teeth are also needed for proper development of a child's speech and chewing of food.

Primary Tooth Eruption

- A general rule of thumb is that for every 6 months of life, approximately 4 teeth will erupt
- Girls generally precede boys in tooth eruption
- Lower teeth usually erupt before upper teeth
- Teeth in both jaws usually erupt in pairs — one on the right and one on the left

- Primary teeth are smaller in size and whiter in color than the permanent teeth that will follow
- By the time a child is 2–3 years of age, all primary teeth should have erupted.

Shortly after age 4, the jaw and facial bones of the child begin to grow, creating spaces between the primary teeth. This is a perfectly natural growth process that provides the necessary space for the larger permanent teeth to emerge. Between the ages of 6 and 12, a mixture of both primary teeth and permanent teeth reside in the mouth.

While it is true that primary teeth are only in the mouth a short period of time, they play a vital role in the following ways:
- They reserve space for their permanent counterparts
- They give the face its normal appearance
- They aid in the development of clear speech
- They help attain good nutrition (missing or decayed teeth make it difficult to chew causing children to reject foods)

- They help give a healthy start to the permanent teeth (decay and infection in baby teeth can cause dark spots on the permanent teeth developing beneath it).

ROLE

The first permanent tooth usually appears in the mouth at around six years of age, and the mouth will then be in a transition time with both primary (deciduous teeth) teeth and permanent teeth during the mixed dentition period until the last primary tooth is lost or shed. The first of the permanent teeth to erupt are the permanent first molars, right behind the last 'milk' molars of the primary dentition. These first permanent molars are important for the correct development of a permanent dentition. Up to the age of thirteen years, twenty-eight of the thirty-two permanent teeth will appear.

The full permanent dentition is completed much later during the permanent dentition period. The four last permanent teeth, the third molars, usually appear between the ages of 17 and 25 years; they are considered wisdom teeth (Table 10.2).

Eruption Sequence of Permanent Dentition

Permanent teeth	Calcification begins	Appearance
Lower incisors	3–4 mo	6–8 yrs
Upper incisors	4–5 mo	7–9 yrs
Lower canines	4–5 mo	9–10 yrs
Upper canines	5–6 mo	11–12 yrs
Premolars	1.5–2.5 yrs	10–12 yrs
1st molars	Birth	6–7 yrs
2nd molars	2.5–3 yrs	11–13 yrs
3rd molars	7–10 yrs	17–21 yrs

Table 10.2: *Emergence ages (years) of permanent teeth in children of different ethnic origins*

Permanent teeth	US (boys)	Nothern Ireland (both)	Gambia (girls)	Chinese (girls)	Chandigarh (both)	Delhi (girls)	Delhi[a] (boys)
Incisor							
Central							
Upper	7.5	7.1	7.1	7.2	6.9	6.8	7.0
Lower	6.6	6.3	6.1	6.1	6.3	6.3	6.1
Lateral							
Upper	8.7	8.2	8.1	8.3	7.5	7.8	8.0
Lower	7.8	7.4	7.1	7.2	7.6	7.4	7.2
Canine							
Upper	11.5	11.2	10.5	10.4	11.0	10.7	9.9
Lower	10.9	10.3	9.7	9.6	11.0	9.5	9.7
Premolar							
First							
Upper	10.8	10.6	9.8	9.5	10.5	9.9	9.7
Lower	11.1	10.5	9.9	9.8	10.5	9.7	9.7
Second							
Upper	11.7	11.4	10.6	10.4	11.2	10.4	10.6
Lower	12.0	11.4	10.7	10.7	11.5	10.4	10.8
Molar							
First							
Upper	6.8	6.4	5.8	6.2	5.9	5.9	5.7
Lower	6.7	6.3	5.5	5.0	6.3	5.7	5.7
Second							
Upper	12.7	12.1	11.2	12	11.5	11.9	11.6
Lower	12.3	11.9	10.9	11.3	11.8	11.7	11.3

[a]Agarwal et al 2004

PATHOLOGY

It is possible to have extra, or 'super-numerary', teeth. This phenomenon is called hyperdontia and is often erroneously referred to as 'a third set of teeth'. These teeth may erupt into the mouth or remain impacted in the bone. Hyperdontia is often associated with syndromes such as cleft lip and palate, trichorhinophalangeal syndrome, cleidocranial dysplasia, and Gardner's syndrome.

SUGGESTED READING

1. Agarwal KN, Gupta R, Faridi MMA, et al. Permanent dentition in Delhi boys of age 5–14 years. Indian Pediatr 2004;41:1031–4.
2. Agarwal KN, Narula S, Faridi MMA, et al. Deciduous dentition and enamel defects. IBID 2004;40:124–9.

11

Growth Curves for 5–18 Years Childern Using LMS Method

KN Agarwal, AK Bansal, DK Agarwal

The height and weight data (Agarwal et al)[1,2,3] growth curves were drawn by fitting polynomic trend using non-linear regression (Chapter 1; Figs 1.9 and 1.10). Recent school children data[4,5] are analyzed to study secular trend, these are analyzed by using LMS method of Cole et al[6]. Therefore it became necessary to reanalyze the earlier data[1] and redraw growth curves using LMS method to make these comparable.

RESULTS

For reference the earlier growth curves drawn by fitting polynomic trend using non-linear regression are shown in Figs 1.9 and 1.10 (Chapter 1; pages 14 and 15). The LMS method calculated growth curves are presented in Figs 11.1 and 11.2: LMS-1 and LMS-2. The growth data tables with LMS method are also given in Tables 11.1 to 11.4.

Table 11.1: *Height percentiles for boys using LMS method (Agarwal et al, Indian Pediatr 1992)*

Age (Yrs)	3rd	10th	25th	50th	75th	90th	97th
5	100.49	103.15	106.02	109.46	113.17	116.79	120.65
6	105.16	107.98	111.02	114.65	118.57	122.38	126.43
7	109.85	112.83	116.05	119.88	124.01	128.01	132.25
8	114.52	117.68	121.08	125.12	129.46	133.66	138.10
9	119.14	122.48	126.07	130.32	134.88	139.27	143.91
10	123.72	127.25	131.03	135.50	140.27	144.85	149.67
11	128.46	132.18	136.15	140.82	145.78	150.52	155.47
12	133.68	137.59	141.73	146.57	151.68	156.52	161.54
13	139.51	143.56	147.82	152.77	157.94	162.80	167.79
14	145.50	149.59	153.86	158.79	163.90	168.68	173.55
15	150.79	154.77	158.93	163.70	168.64	173.22	177.89
16	154.98	158.77	162.72	167.24	171.90	176.23	180.61
17	158.41	161.96	165.66	169.88	174.23	178.25	182.32
18	161.59	164.89	168.32	172.22	176.23	179.94	183.69

The differences in height and weight in the polynomic trend and B-LMS method for height (cm) and weight (kg) are given in Table 11.5, showing marginal differences.

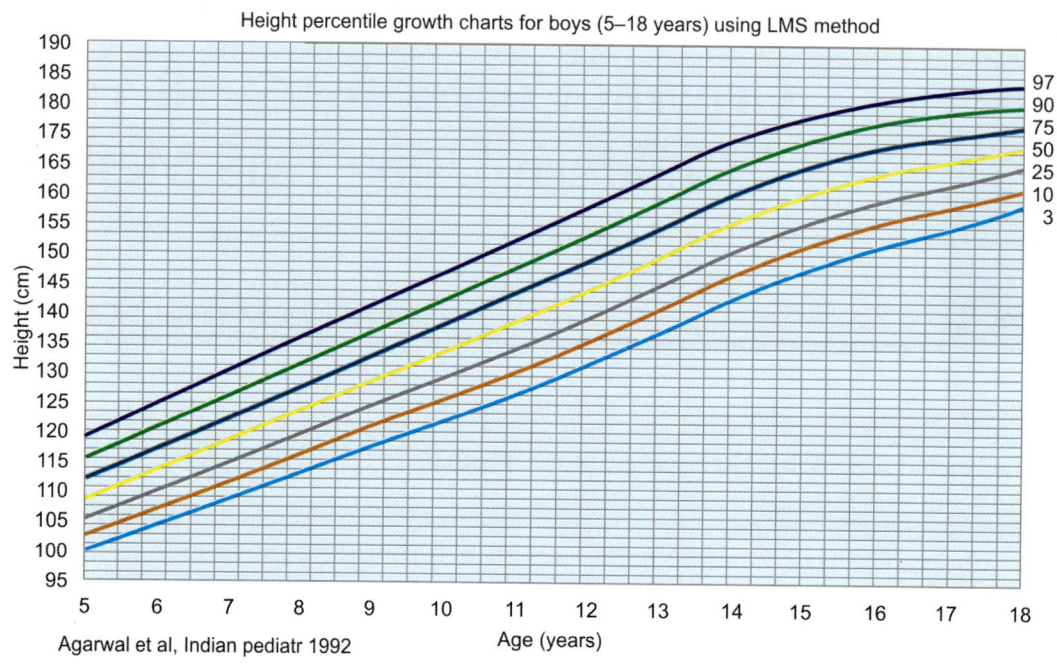

Agarwal et al, Indian pediatr 1992

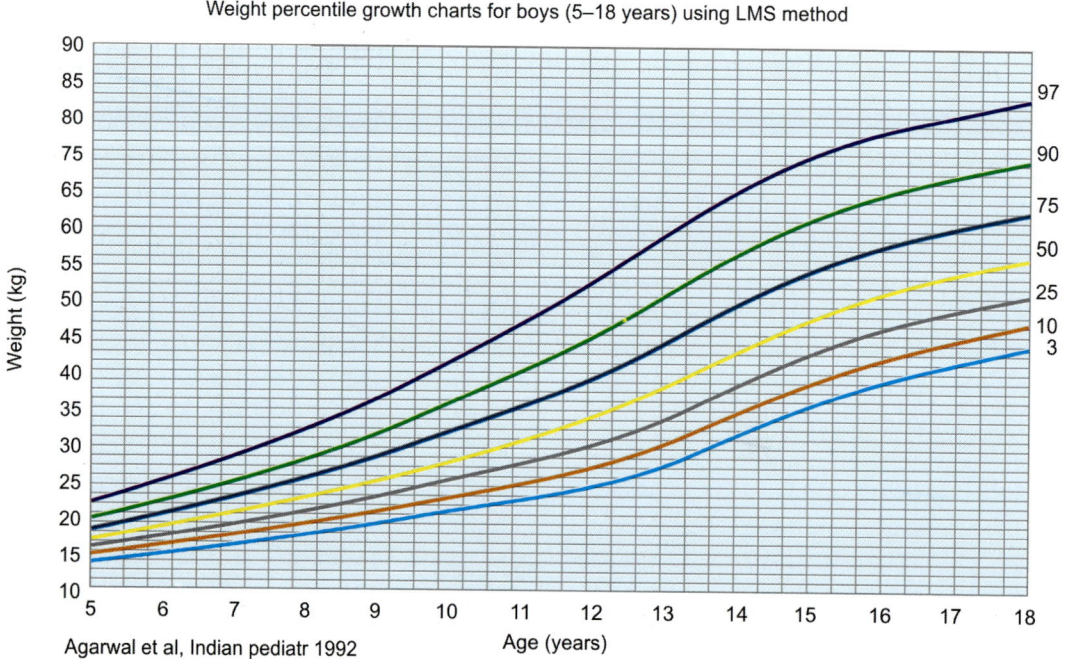

Agarwal et al, Indian pediatr 1992

Fig. 11.1: *LMS-1: calculated growth curves for boys*

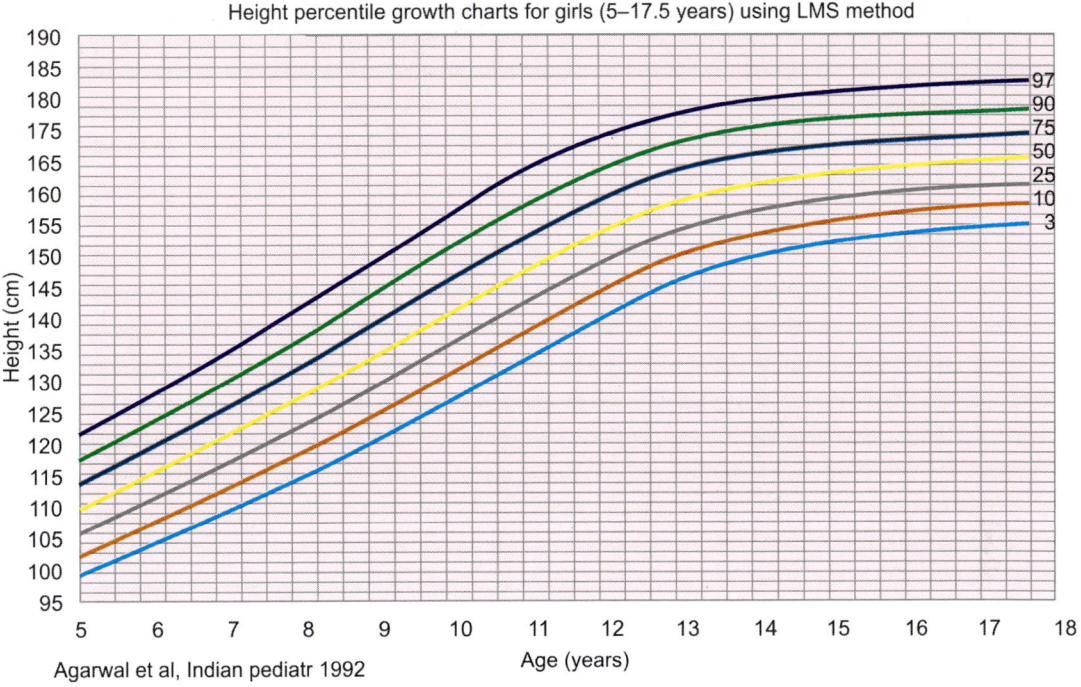

Height percentile growth charts for girls (5–17.5 years) using LMS method

Agarwal et al, Indian pediatr 1992

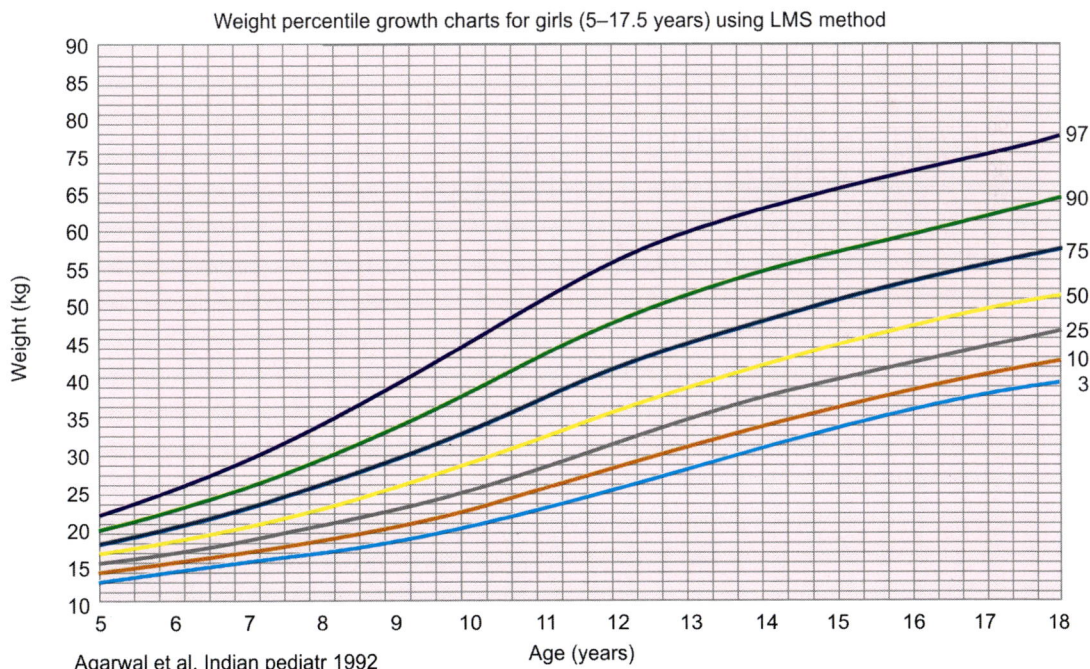

Weight percentile growth charts for girls (5–17.5 years) using LMS method

Agarwal et al, Indian pediatr 1992

Fig. 11.2: *LMS-2: calculated growth curves for girls*

Table 11.2: *Weight percentiles for boys using LMS method (Agarwal et al, Indian Pediatr 1992)*

Age (Yrs)	3rd	10th	25th	50th	75th	90th	97th
5	14.04	14.88	15.86	17.15	18.72	20.45	22.58
6	15.43	16.43	17.61	19.16	21.07	23.20	25.86
7	16.78	17.96	19.35	21.22	23.52	26.11	29.38
8	18.16	19.55	21.21	23.43	26.20	29.35	33.37
9	19.66	21.29	23.24	25.89	29.21	33.03	37.92
10	21.27	23.18	25.49	28.62	32.60	37.19	43.13
11	23.01	25.25	27.95	31.65	36.35	41.81	48.88
12	25.27	27.86	31.00	35.30	40.76	47.08	55.22
13	28.43	31.38	34.95	39.81	45.95	53.00	61.99
14	32.52	35.77	39.67	44.93	51.51	58.98	68.37
15	36.76	40.18	44.25	49.69	56.44	64.01	73.42
16	40.30	43.78	47.91	53.40	60.14	67.67	76.95
17	43.11	46.61	50.75	56.23	62.93	70.38	79.53
18	45.56	49.08	53.24	58.72	65.42	72.85	81.97

Table 11.3: *Height percentiles for girls using LMS method (Agarwal et al, Indian Pediatr 1992)*

Age (Yrs)	3rd	10th	25th	50th	75th	90th	97th
5	100.08	102.80	105.67	109.03	112.56	115.90	119.35
6	104.57	107.45	110.51	114.07	117.81	121.35	125.01
7	109.11	112.18	115.43	119.21	123.18	126.92	130.80
8	113.85	117.13	120.60	124.62	128.84	132.82	136.93
9	118.86	122.39	126.10	130.41	134.92	139.15	143.50
10	124.25	128.01	131.96	136.51	141.26	145.71	150.27
11	130.21	133.99	137.97	142.54	147.30	151.75	156.29
12	136.21	139.78	143.51	147.81	152.28	156.46	160.73
13	141.06	144.34	147.78	151.76	155.91	159.79	163.78
14	144.11	147.19	150.43	154.20	158.15	161.86	165.70
15	145.89	148.82	151.91	155.53	159.34	162.95	166.71
16	147.05	149.84	152.80	156.29	159.98	163.50	167.19
17	147.89	150.57	153.44	156.82	160.43	163.89	167.54
17.5	148.28	150.91	153.73	157.05	160.62	164.05	167.67

Table 11.4: *Weight percentiles for girls using LMS method (Agarwal et al, Indian Pediatr 1992)*

Age (Yrs)	3rd	10th	25th	50th	75th	90th	97th
5	13.16	14.07	15.13	16.52	18.20	20.03	22.25
6	14.41	15.53	16.85	18.59	20.73	23.11	26.04
7	15.70	17.06	18.67	20.83	23.51	26.55	30.35
8	17.19	18.82	20.78	23.44	26.78	30.60	35.46
9	18.97	20.93	23.29	26.52	30.62	35.35	41.41

contd.

Table 11.4: *Weight percentiles for girls using LMS method (Agarwal et al, Indian Pediatr 1992) (contd.)*

Age (Yrs)	3rd	10th	25th	50th	75th	90th	97th
10	21.16	23.47	26.27	30.11	35.00	40.65	47.92
11	23.78	26.42	29.63	34.02	39.60	46.02	54.23
12	26.73	29.64	33.15	37.93	43.95	50.84	59.55
13	29.84	32.90	36.58	41.54	47.72	54.73	63.50
14	32.81	35.94	39.68	44.68	50.87	57.80	66.41
15	35.39	38.58	42.36	47.40	53.59	60.50	69.01
16	37.62	40.88	44.73	49.84	56.11	63.06	71.61
17	39.69	43.02	46.96	52.16	58.53	65.59	74.24
17.5	40.70	44.07	48.05	53.31	59.74	66.85	75.57

Table 11.5. Showing differences observed for A-Polynomic trend and B-LMS method for height (cm) and weight (kg) at 18 and 17.5 years of age for boys and girls respectively.

								Table 11.5	
Percentiles	Height A		Height B		Weight A		Weight B	
	Boys	(Girls)	Boys	(Girls)	Boys	(Girls)	Boys	(Girls)
3rd	160.7	(149.1)	161.6	(148.2)	45.9	(38.5)	45.7	(40.7)
50th	168.9	(158.3)	172.2	(157.1)	57.1	(48.6)	58.7	(53.3)
97th	180.1	(168.8)	183.7	(167.7)	82.6	(76.0)	82.0	(75.6)

REFERENCES

1. Agarwal DK, Agarwal KN, Upadhyay SK, et al. Physical and sexual growth pattern of affluent Indian children from 5 to 18 years of age. Indian Pediatr 1992;29:1203–82.

2. Agarwal KN, Saxena A, Bansal AK, et al. Physical growth assessment in adolescence. Indian Pediatr 2001;38:1217–35.

3. Khadgawat R, Dabadghao P, Mehrotra RN, et al. Growth charts for evaluation of Indian children. Indian Pediatr 1998;35:859–65.

4. Khadilkar VV, Khadilkar AV, Cole TJ, et al. Cross-sectional Growth Curves for Height, Weight and Body Mass Index for Affluent Indian Children, 2007. Indian Pediatr 2009;46:477–89.

5. Marwaha RK, Tandon N, Ganie MA, et al. Nationwide reference data for height, weight and body mass index of Indian schoolchildren. Natl Med J India 2011;24:269–77.

6. Cole TJ, Green PJ. Smoothing reference centile curves: The LMS method and penalized likelihood. Stat Med 1992;11:1305–19.

12

Z-scores for Calculation of Malnutrition < 5 Years of Age

AK Bansal, DK Agarwal, KN Agarwal

The current use of **WHO growth data** (WHO Multicenteric Growth Reference Study Group: WHO child growth standards based on length/height, and weight and age Acta Pediatr 2006; Suppl 450:76–85.) **as standard** over diagnoses–Underweight and Stunted children. These Z-score curves can be seen in WHO website (http://www.cdc.gov/growthcharts/who_charts.htm#).

There is need to use Indian reference growth data (as reference data show how children are growing in time and place) to assess the exact extent of undernutrition and stunting, see Z-score growth charts below using growth data of Agarwal, and Agarwal.

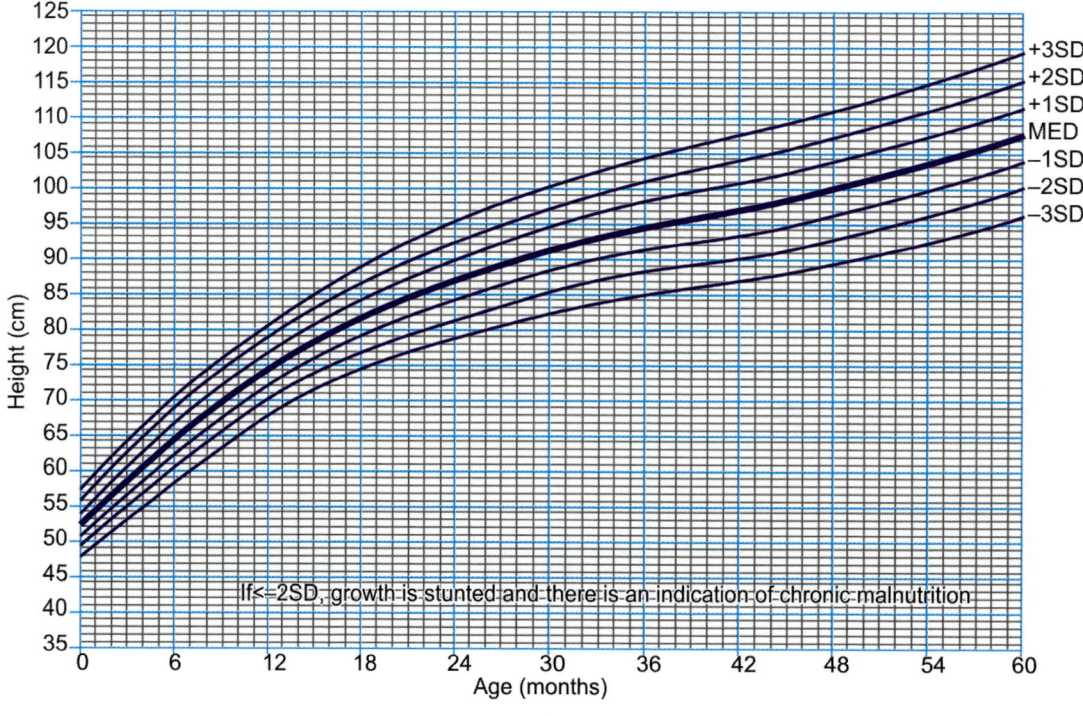

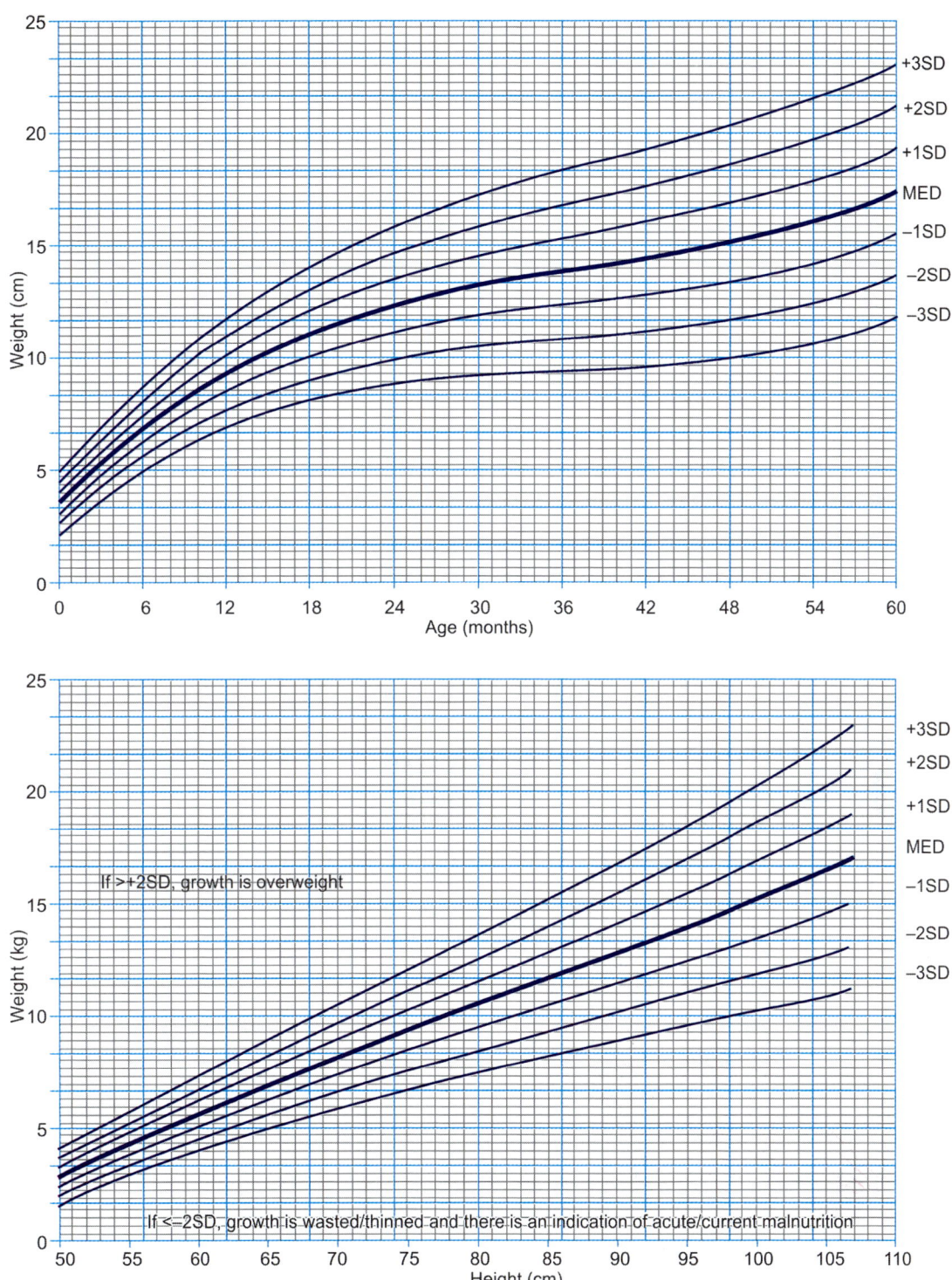

Physical growth in affluent Indian children (birth–6 years) Indian Pediatr 1994;31:377–413.

i. Z-scores for weight-for-age
ii. Z-scores for height-for-age, and
iii. Z-scores for weight-for-height

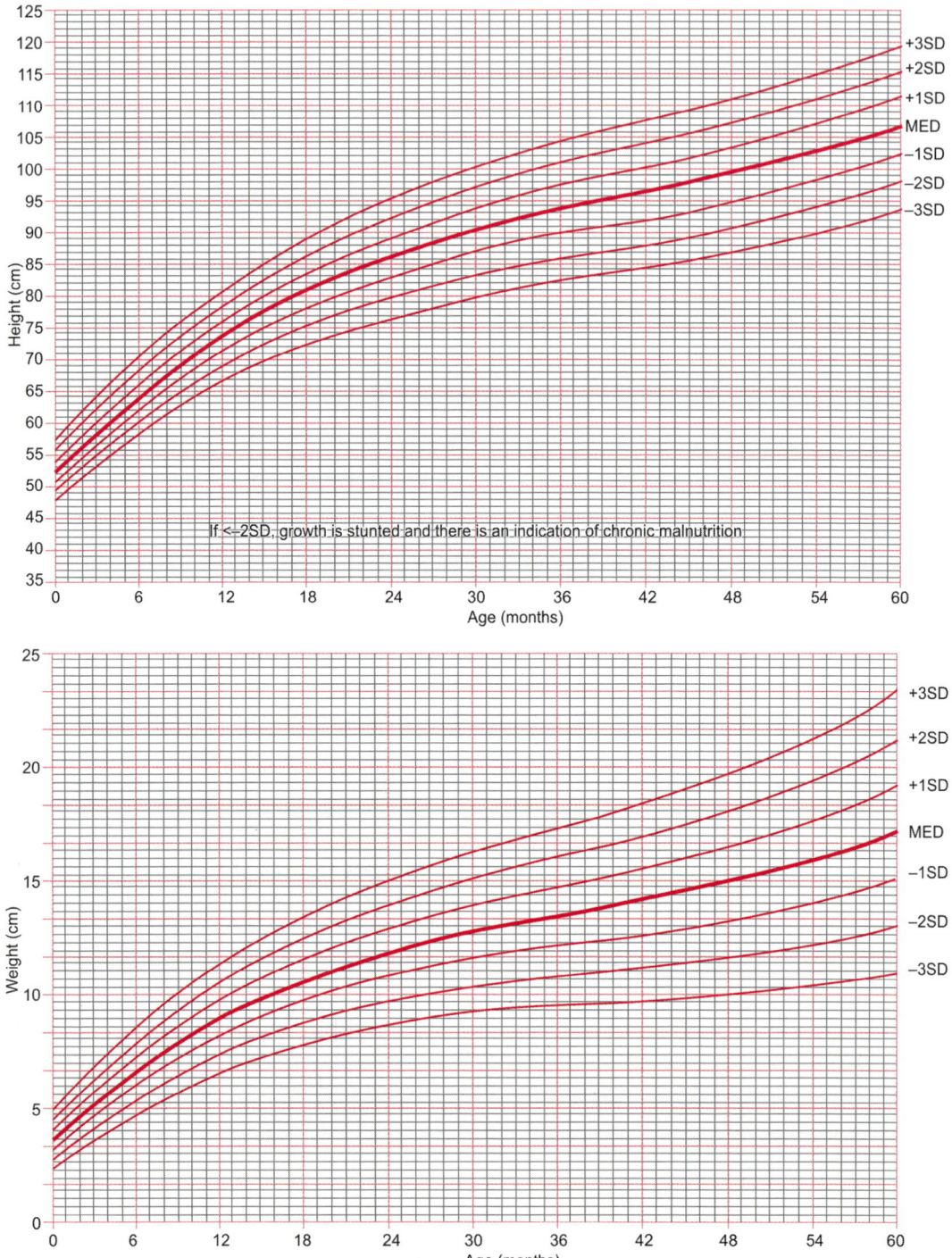

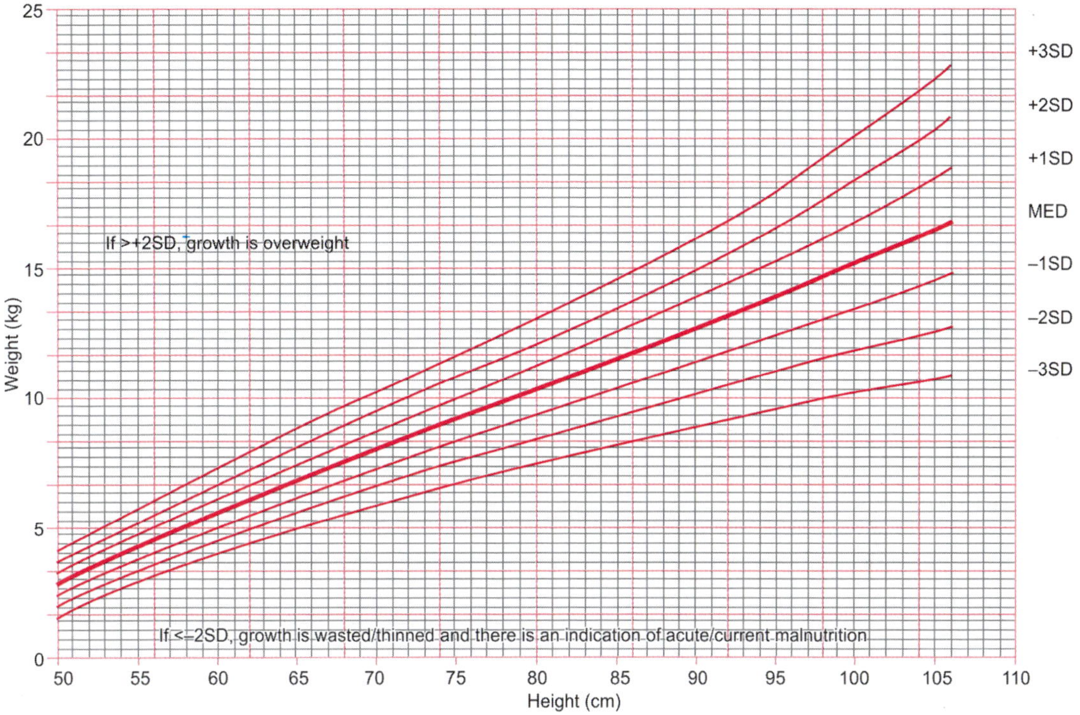

The chart shows Weight (kg) on the y-axis (0 to 25) versus Height (cm) on the x-axis (50 to 110), with curves for +3SD, +2SD, +1SD, MED, -1SD, -2SD, -3SD.

If >+2SD, growth is overweight

If <-2SD, growth is wasted/thinned and there is an indication of acute/current malnutrition

Z-score growth charts for girls aged 0–5 years (Agarwal and Agarwal 1994)

i. Z-scores for weight-for-age
ii. Z-scores for height-for-age, and
iii. Z-scores for weight-for-height

Comparison between the Indian Affluent and the WHO SD Scores

The SD scores of Indian affluent boys <5 years are compared with the WHO SD scores. From these figures, it is evident that the shape and pattern followed is almost similar for both the data. The WHO curves being higher than the Indian.

We have not plotted all the Z-score values in a single chart as it makes it clumsy. Instead, we have plotted individual charts for all seven sets of Z-scores, i.e. –3SD, –2SD, –1SD, median, +1SD, +2SD, +3SD for height in boys.

Height-for-age (boys)

From these figures, it is seen that the shape or pattern followed is almost similar for both the data. In the later age say from

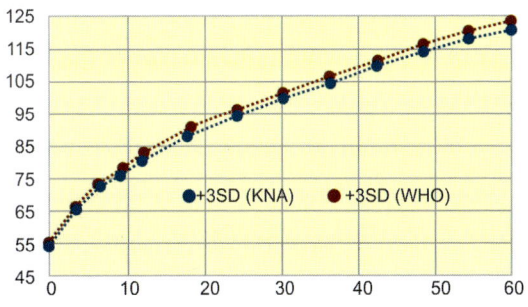

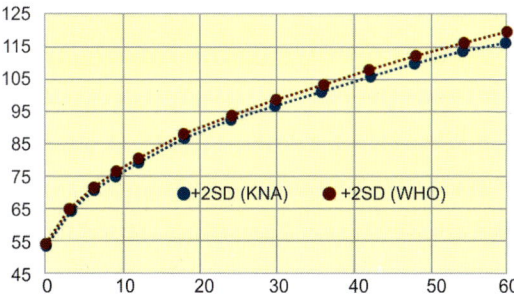

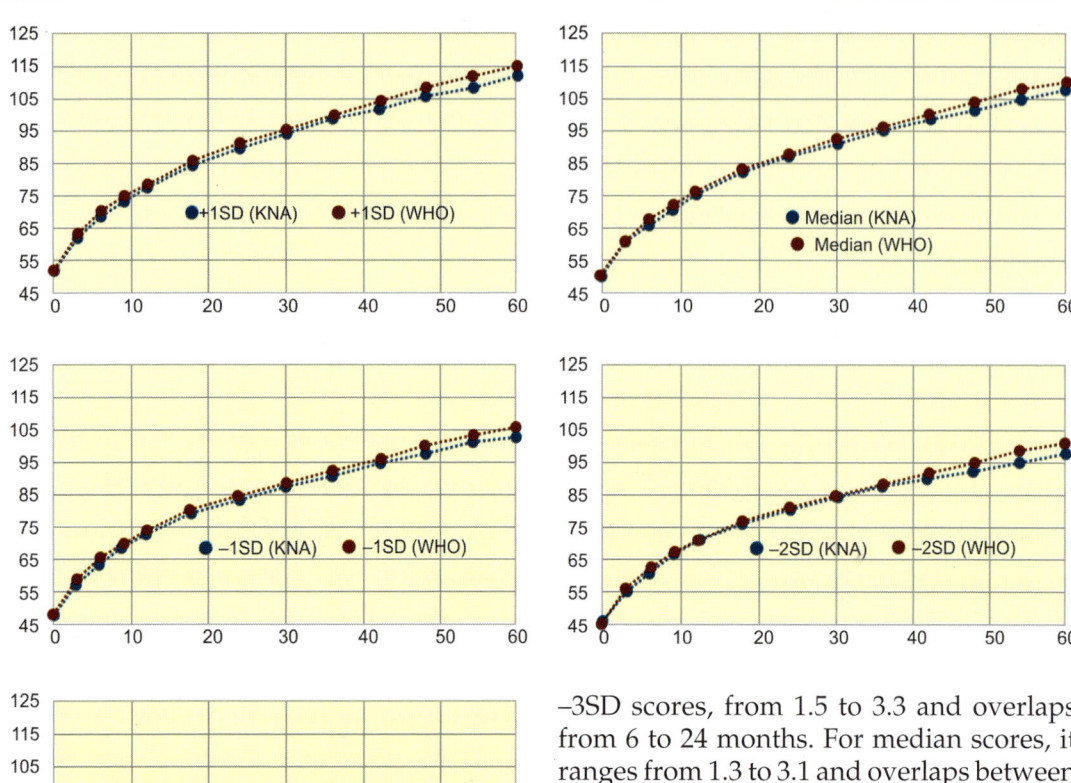

3 to 5 years, the difference in Z-scores is more as compared to 0 to 3 years of age. For +3SD scores, it ranges from 1.3 to 2.9 and overlaps between 0 and 6 months, for –3SD scores, from 1.5 to 3.3 and overlaps from 6 to 24 months. For median scores, it ranges from 1.3 to 3.1 and overlaps between 0 and 3 months. In the initial age, i.e. from 0 to 3 years, the shape follows the similar pattern and distribution of the attained height.

From these figures, it is clear that the use of WHO Z-score curves will over diagnose undernutrition/stunting and miss the overweight children. It is suggested that the Z-score curves developed on Indian affluent be used as reference for our children.

Appendix

DEVELOPMENT OF GROWTH CURVES NCHS/WHO/INDIAN

Data Selected in Development of NCHS–WHO (Reference) and Indian (Standard) Growth Curves

To understand science of development of growth curves, it will be of interest to learn from the CDC growth charts, released in May 2000. These consist of revised versions of the growth charts developed by the NCHS in 1977 with the addition of the BMI-for-age charts. It is important to note CDC–NCHS growth charts designed in 2000 did not collect new primary data. Infact they used National studies conducted at various places and time. As the development of growth curves by collecting primary data is very expensive and time consuming.

Table 1: *NCHS/CDC recommends growth curves (1963 to 1994 data)*

- Weight, length, head circumference and weight-for-height curve birth to 2 years of age for both sexes.
- For 2–20 years of age—height and BMI curves for both sexes.
- In adolescence BMI, height- and weight-for-age are different in sexual maturity stages and relate to sexual development (Agarwal et al, Indian Pediatrics, 2001).

Data Sets Used (1963–94) for Preparing the WHO/CDC Growth Curves for Birth to 2 Years of Age

- NHANES-III (1988–94)—weight, length and head circumference at 2 months of age
- NHANES-II (1976–80)—data beginning of 6 months of age
- NHANES-I (1971–74)—data beginning of 12 months of age
- National survey did not collect measurements between birth to 2 months of age
  a. **Birth data** US vital statistics—length and weight for infant from Missouri and Wisconsin birth certificates.
  b. **Length data** 0.5–4.5 mo of age in the *Pediatric Nutrition Surveillance System* (50% breastfed, 33% fed longer than 3 months). Breastfed grow less rapidly than the formula fed until 12 months of age
  c. **Head circumference** at birth from *Fels Research Institute* longitudinal data (1929–1975—mainly formula fed).

Growth charts excluded values for very low birth weight babies.

Growth charts for aged 2–20 years of age (Nationally representative from 5 national surveys covering all ethnic groups)

Developed using data from various sources:

- NHES—Cycle II, included weight and stature data from children and adolescents from 6 through 11 years of age.
- Data from Cycle III of NHES represented children and adolescents from 12 through 17 years of age.
- NHANES I, II, and III provided weight and stature data for children from 2 to 20 years of age. However, the NHANES III weight measures (1988–94) were excluded for children 6 years of age and older (prevalence of obesity had significantly increased as compared to 1976–84 data).

The CDC/NCHS curves do not take in account the racial differences. Afro-Americans are taller than the European-American, while Asian-Americans are shorter. If an Afro-American child is less than 50th centile of the NCHS height, he/she is abnormal and should be investigated.

The recent WHO (April 2006) multinational growth curves for boys and girls 0–60 months are different than the NCHS (CDC) reference discussed above:

a. Populations used are different

b. Methodologies are more advanced in the new WHO reference.

The differences are particularly important in infancy. Stunting will be greater throughout childhood when assessed using the new WHO standards compared to the NCHS reference. The new WHO curves will give substantial increase in rates of underweight during the first half of infancy and a decrease thereafter. For wasting, the main difference is during infancy when wasting rates are higher using the new WHO standards. With respect to overweight, use the new WHO standards will result in a greater prevalence that will vary by age, sex and nutritional status of the index population. (www.who.int/childgrowth.) (*see* Tables 1 to 6 and Figs 1 and 2).

Indian Growth Standards (Tables 7 to 25)

Development of local growth standard

a. Needs affluent/elite children with unconstrained growth,

b. Geographically well distributed,

c. At least 200 children at each age and sex point,

d. Measurements taken by standard tools,

e. The measurements should follow Gaussian distribution

f. Children should have normal expected growth velocity and g) possibly by the same team. Ideally study population should not be in secular growth trend.

The above mentioned WHO criteria were followed in developing the growth standard for the Indian children. Indian affluent children growth data for 0–6 years were collected as a semi-longitudinal study in a cohort of 1011 boys and 874 girls, using the cross-linked method. For this age group data were collected from Bangalore (Bengaluru), Calcutta (Kolkata), Kota, Delhi, Ludhiana and Varanasi (Nutrition Foundation of India Sci Rep 11, 1991). Healthy affluent babies with birth weight >2500 g with no abnormality were selected. This method allowed velocity calculations. For children in the age group of 6–18 years, data were collected from Bombay (Mumbai), Kolkata, Dhanbad, Dehradun, Delhi, Lucknow, Madras (Chennai), Shimla and Udaipur. This was a cross-sectional study on 12893 boys and 10941 girls (Indian Council of Medical Research Study). The growth curves were constructed using cubic spline method (*see* Chapter 11). The 50th centile height measure on the pooled data variation was <3% as compared to the 50th centile derived from each sample (city of study).

The height for Indian affluent children was lower by 1 cm at 24 months, 3 cm at 5–6 years and 7 cm at 17–18 years of age as compared to the CDC growth data. The 50th centile for 18 years of age of the Indian boy corresponded between 10th and 25th centile of the CDC. However, the affluent Indian child is similar to other Asian children for linear growth pattern (Chinese, Japanese, Korean, Taiwanese). Maximum height difference being 1 cm at 17 years of age.

In 2002, Agarwal et al reassessed 2000 Delhi affluent school children. There was no secular trend. However, obesity had increased from <1% to 10% after an interval of 12 years. Virani 2005 reported on Aurobindo Ashram children from Pondicherry (puducherry) followed for 40 years. The secular trend was not observed for last 20 years, but Indian are shorter than the Europeans.

A systemic review worldwide variations in human growth and the WHO [MGRS] growth standards. BMJ open 2014;4:1–11; March 13th 2014 concludes use of National growh standards after analyzing data from 55 countries (Natale and Rajagopalan).

Growth Data Contents

1. NCHS and WHO growth data are shown in Tables 1 to 6 and Figs 1 and 2.
2. Indian affluent children data are shown in Tables 7 to 25.

Table 1: *Weight (kg) by age of boys and girls aged 0–5 years (WHO)*

Age		Percentiles for boys							Percentiles for girls						
Years	months	3rd	5th	25th	50th	75th	95th	97th	3rd	5th	25th	50th	75th	95th	97th
0	0	2.5	2.6	3.0	3.3	3.7	4.2	4.3	2.4	2.5	2.9	3.2	3.6	4.0	4.2
0	3	5.1	5.2	5.9	6.4	6.9	7.7	7.9	4.6	4.7	5.4	5.8	6.4	7.2	7.4
0	6	6.4	6.6	7.4	7.9	8.5	9.5	9.7	5.8	6.0	6.7	7.3	7.9	8.9	9.2
0	9	7.2	7.4	8.3	8.9	9.6	10.6	10.9	6.6	6.8	7.6	8.2	8.9	10.1	10.4
1	0	7.8	8.1	9.0	9.6	10.4	11.5	11.8	7.1	7.3	8.2	8.9	9.7	11.0	11.3
1	3	8.4	8.6	9.6	10.3	11.1	12.3	12.7	7.7	7.9	8.8	9.6	10.4	11.8	12.2
1	6	8.9	9.1	10.1	10.9	11.8	13.1	13.5	8.2	8.4	9.4	10.2	11.1	12.6	13.0
1	9	9.3	9.6	10.7	11.5	12.5	13.9	14.3	8.7	8.9	10.0	10.9	11.8	13.4	13.8
2	0	9.8	10.1	11.3	12.2	13.1	14.7	15.1	9.2	9.4	10.6	11.5	12.5	14.2	14.6
2	6	10.7	11.0	12.3	13.3	14.4	16.2	16.6	10.1	10.4	11.7	12.7	13.8	15.7	16.2
3	0	11.4	11.8	13.2	14.3	15.6	17.5	18.0	11.0	11.3	12.7	13.9	15.1	17.3	17.8
3	6	12.2	12.5	14.1	15.3	16.7	18.9	19.4	11.8	12.1	13.7	15.0	16.4	18.8	19.5
4	0	12.9	13.3	15.0	16.3	17.8	20.2	20.9	12.5	12.9	14.7	16.1	17.7	20.4	21.1
4	6	13.6	14.0	15.9	17.3	19.0	21.6	22.3	13.2	13.7	15.6	17.2	18.9	22.0	22.8
5	0	14.3	14.7	16.7	18.3	20.1	23.0	23.8	14.0	14.4	16.5	18.2	20.2	23.5	24.4

Table 2: *Weight (kg) by age of boys and girls aged 5–10 years (NCHS)*

Age (years)				Percentiles					
	3rd	5th	15th	25th	50th	75th	85th	95th	97th
Boys									
5	14.3	14.7		16.7	18.3	20.1		23.0	23.8
6	16.1	16.6	17.9	18.8	20.5	22.5	23.6	25.8	26.7
7	17.9	18.4	19.9	20.9	22.9	25.2	26.5	29.1	30.1
8	19.8	20.4	22.0	23.1	25.4	28.1	29.7	32.7	34.0
9	21.6	22.3	24.2	25.4	28.1	31.3	33.2	36.9	38.6
10	23.6	24.4	26.6	28.0	31.2	34.9	37.3	41.9	43.9
Girls									
5	14.0	14.4		16.5	18.2	20.2		23.5	24.4
6	15.5	16.0	17.4	18.3	20.2	22.4	23.7	26.2	27.3
7	17.0	17.6	19.2	20.2	22.4	24.9	26.5	29.5	30.8
8	18.9	19.5	21.3	22.5	25.0	28.0	29.8	33.4	34.9
9	21.1	21.8	23.9	25.3	28.2	31.7	33.9	38.1	40.0
10	23.7	24.5	26.9	28.5	31.9	35.9	38.5	43.5	45.7

Table 3a: *Length (cm) by age of boys and girls aged 0–2 years (WHO)*

Age	Percentiles for boys							Percentiles for girls						
(months)	3rd	5th	25th	50th	75th	95th	97th	3rd	5th	25th	50th	75th	95th	97th
0	14.3	46.8	48.6	49.9	51.2	53.0	53.4	45.6	46.1	47.9	49.1	50.4	52.2	52.7
3	57.6	58.1	60.1	61.4	62.8	64.8	65.3	55.8	56.3	58.4	59.8	61.2	63.3	63.8
6	63.6	64.1	66.2	67.6	69.1	71.1	71.6	61.5	62.0	64.2	65.7	67.3	69.5	70.0
9	67.7	68.3	70.5	72.0	73.5	75.7	76.2	65.9	66.2	68.5	70.1	71.8	74.1	74.7
12	71.3	71.8	74.1	75.7	77.4	79.7	80.2	69.2	69.8	72.3	74.0	75.8	78.3	78.9
15	74.4	75.0	77.4	79.1	80.9	83.3	83.9	72.4	73.0	75.7	77.5	79.4	82.0	82.7
18	77.2	77.8	80.4	82.3	84.1	86.7	87.3	75.2	75.9	78.7	80.7	82.7	85.5	86.2
21	79.7	80.4	83.2	85.1	87.1	89.9	90.5	77.9	78.6	81.6	83.7	85.7	88.7	89.4
24	82.1	82.8	85.8	87.8	89.9	92.8	93.6	80.3	81.1	84.2	86.4	88.6	91.7	92.5

Table 3b: *Height-for-age for boys and girls 2–5 years of age (WHO)*

Age		Percentiles for boys							Percentiles for girls						
Years	months	3rd	5th	25th	50th	75th	95th	97th	3rd	5th	25th	50th	75th	95th	97th
2	0	81.4	82.1	85.1	87.1	89.2	92.1	92.9	79.6	80.4	83.5	85.7	87.9	91.0	91.8
2	6	85.5	86.3	89.6	91.9	94.2	97.5	98.3	84.0	84.9	88.3	90.7	93.1	96.5	97.3
3	0	89.1	90.0	93.6	96.1	98.6	102.2	103.1	87.9	88.8	92.5	95.1	97.6	101.3	102.2
3	6	92.4	93.3	97.2	99.9	102.5	106.4	107.3	91.4	92.4	96.3	99.0	101.8	105.7	106.7
4	0	95.4	96.4	100.5	103.3	106.2	110.2	111.2	94.6	95.6	99.8	102.7	105.6	109.8	110.8
4	6	98.4	99.4	103.7	106.7	109.6	113.9	115.0	97.6	98.7	103.1	106.2	109.2	113.6	114.7
5	0	101.2	102.3	106.8	110.0	113.1	117.6	118.7	100.5	101.6	106.2	109.4	112.6	117.2	118.4

Table 4: *Height (cm) by age of boys and girls aged 5–10 years (NCHS)*

Age (years)					Percentiles					
	3rd	5th	15th	25th	50th	75th	85th	95th	97th
Boys									
5	101.1	102.3		106.8	110.0	113.1		117.6	118.7
6	106.7	107.8	110.8	112.6	116.0	119.3	121.1	124.1	125.2
7	111.8	113.0	116.3	118.2	121.7	125.3	127.2	130.4	131.7
8	116.6	118.0	121.4	123.5	127.3	131.1	133.1	136.6	137.9
9	121.3	122.7	126.3	128.5	132.6	136.6	138.8	142.5	143.9
10	125.8	127.3	131.2	133.5	137.8	142.1	144.4	148.3	149.8
Girls									
5	100.5	101.6		106.2	109.4	112.6		117.2	118.4
6	105.5	106.7	109.8	111.7	115.1	118.6	120.4	123.5	124.8
7	110.5	111.8	115.1	117.1	120.8	124.5	126.5	129.8	131.1
8	115.7	117.0	120.5	122.6	126.6	130.5	132.6	136.1	137.5
9	121.0	122.4	126.2	128.4	132.5	136.6	138.8	142.5	144.0
10	126.6	128.1	132.0	134.3	138.6	143.0	145.3	149.2	150.7

BMI-FOR-AGE BOYS

Birth to 2 years (percentiles)

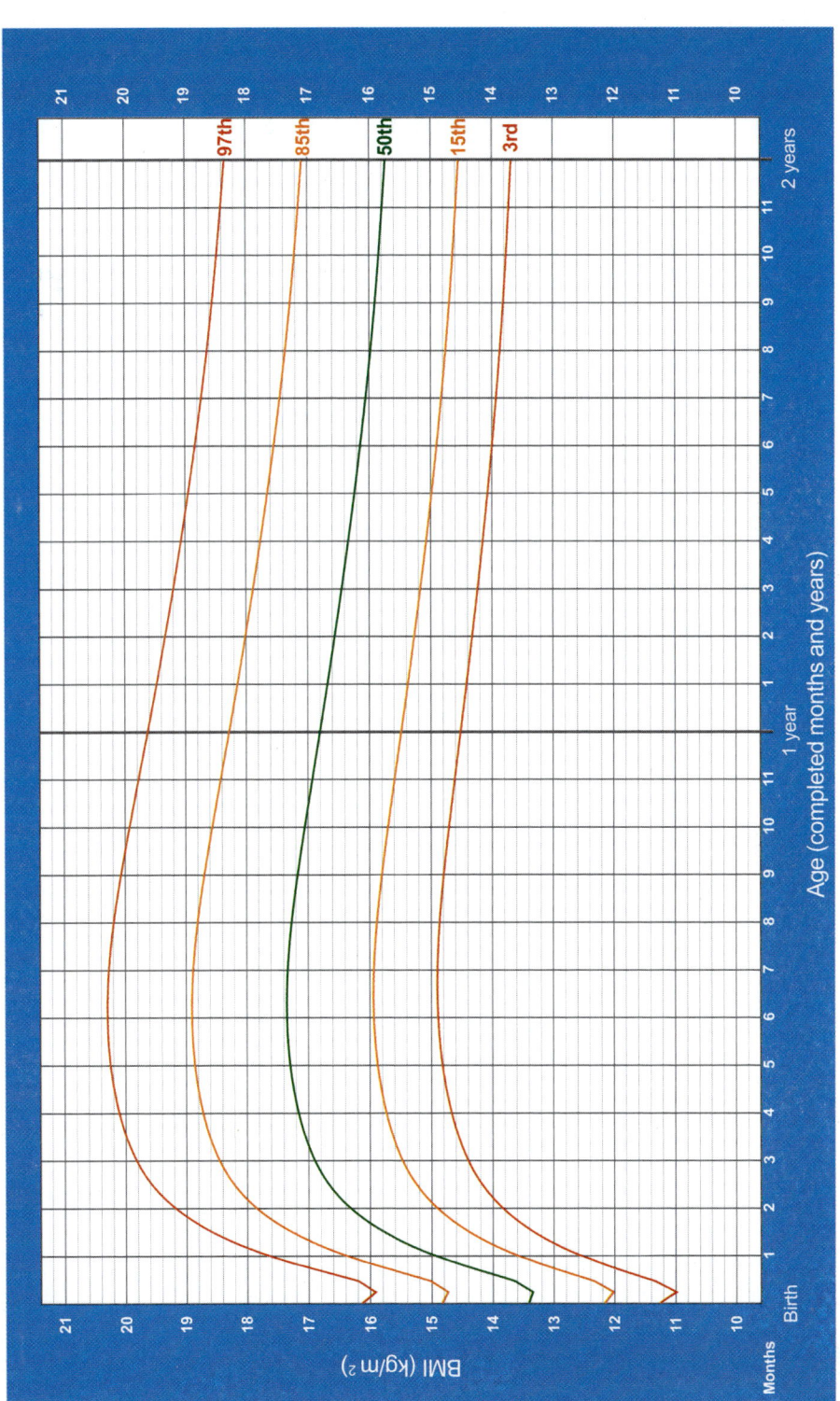

Fig. 1

WHO Child Growth Standards

BMI-FOR-AGE GIRLS

Birth to 2 years (percentiles)

World Health Organization

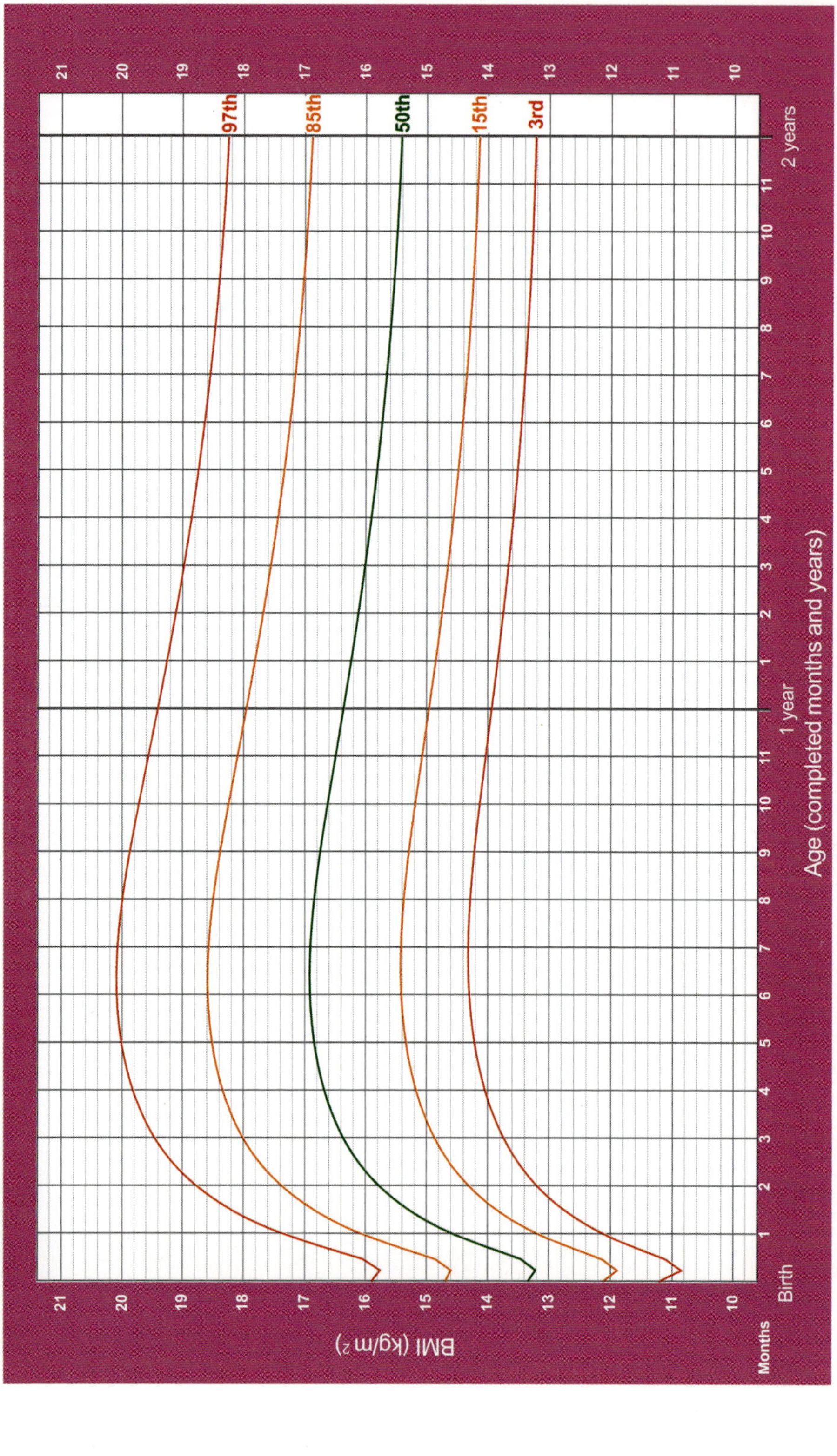

WHO Child Growth Standards

Fig. 2

BMI chart — Months / Birth / 1 year / 2 years axis; Age (completed months and years); BMI (kg/m²); percentile curves labelled 97th, 85th, 50th, 15th, 3rd.

Table 5: *Body mass index (BMI) percentiles for age, boys and girls, 2–20 years (NCHS)*

Boys: Percentiles

Age (yr)	5th	10th	25th	50th	75th	85th	90th	95th
2.0	14.7	15.1	15.7	16.6	17.6	18.2	18.6	19.3
2.5	14.5	14.9	15.5	16.2	17.1	17.7	18.1	18.7
3.0	14.3	14.7	15.3	16.0	16.8	17.3	17.7	18.2
3.5	14.2	14.5	15.1	15.8	16.5	17.1	17.4	18.0
4.0	14.0	14.3	14.9	15.6	16.4	16.9	17.3	17.8
4.5	13.9	14.2	14.8	15.5	16.3	16.8	17.2	17.8
5.0	13.8	14.1	14.7	15.4	16.3	16.8	17.3	17.9
5.5	13.8	14.1	14.6	15.4	16.3	16.9	17.4	18.1
6.0	13.7	14.0	14.6	15.4	16.4	17.0	17.5	18.4
6.5	13.7	14.0	14.6	15.4	16.5	17.2	17.7	18.8
7.0	13.7	14.0	14.7	15.5	16.6	17.4	18.0	19.2
7.5	i3.7	14.1	14.7	15.6	16.8	17.7	18.3	19.6
8.0	13.8	14.1	14.8	15.8	17.1	18.0	18.7	20.1
8.5	13.9	14.2	15.0	16.0	17.3	18.3	19.1	20.6
9.0	14.0	14.3	15.1	16.2	17.6	18.6	19.5	21.1
9.5	14.1	14.5	15.3	16.4	17.9	19.0	19.9	21.6
10.0	14.2	14.6	15.5	16.6	18.2	19.4	20.3	22.2
10.5	14.4	14.8	15.7	16.9	18.6	19.8	20.8	22.7
11.0	14.6	15.0	15.9	17.2	18.9	20.2	21.2	23.2
11.5	14.8	15.2	16.2	17.5	19.3	20.6	21.7	23.7
12.0	15.0	15.5	16.4	17.8	19.7	21.0	22.1	24.2
12.5	15.2	15.7	16.7	18.1	20.1	21.4	22.6	24.7
13.0	15.5	16.0	17.0	18.5	20.4	21.9	23.0	25.2
13.5	15.7	16.3	17.3	18.8	20.8	22.3	23.4	25.6
14.0	16.0	16.5	17.6	19.2	21.2	22.7	23.8	26.0
14.5	16.3	16.8	17.9	19.5	21.6	23.1	24.3	26.5
15.0	16.6	17.1	18.3	19.9	22.0	23.5	24.6	26.8
15.5	16.8	17.4	18.6	20.2	22.4	23.8	25.0	27.2
16.0	17.1	17.7	18.9	20.6	22.7	24.2	25.4	27.6
16.5	17.4	18.0	19.2	20.9	23.1	24.6	25.8	27.9
17.0	17.7	18.3	19.6	21.2	23.4	24.9	26.1	28.3
17.5	18.0	18.6	19.9	21.6	23.8	25.3	26.5	28.6
18.0	18.2	18.9	20.2	21.9	24.1	25.7	26.9	29.0
18.5	18.5	19.2	20.4	22.2	24.5	26.0	27.2	29.3
19.0	18.7	19.4	20.7	22.5	24.8	26.4	27.6	29.7
19.5	18.9	19.6	21.0	22.8	25.1	26.7	28.0	30.2
20.0	19.1	19.8	21.2	23.0	25.4	27.0	28.3	30.6

Contd.

Table 5: *Body mass index (BMI) percentiles for age, boys and girls, 2–20 years (NCHS) (Contd.)*

Girls: Percentiles

Age (yrs)	5th	10th	25th	50th	75th	85th	90th	95th
2.0	14.4	14.8	15 5	16.4	17.4	18.0	18.4	19.1
2.5	14.2	14.5	15.2	16.0	16.9	17.5	17 9	18.6
3.0	14.0	14.3	14.9	15.7	16.6	17.2	17.6	18.3
3.5	13.8	14.2	14.7	15.5	16.4	16.9	17.4	18.1
4.0	13.7	14.0	14 6	15.3	16.2	16.8	17.3	18.0
4.5	13.6	13.9	145	15.2	16.1	16.8	17.2	18.1
5.0	13.5	13.8	14.4	15.2	16.1	16.8	17.3	18.3
5.5	13.5	13.8	14.4	15.2	16.2	16.9	17.5	18.5
6.0	13.4	13.7	14.4	15.2	16.3	17.1	17.7	18.8
6.5	13.4	13.8	14 4	15.3	16.5	17.3	18.0	19.2
7.0	13 4	13.8	14.5	15.5	16.7	17.6	18.3	19.7
7.5	13.5	13.9	14.6	15.6	17.0	18.0	18.7	20.2
8.0	13.5	13.9	14.7	15.8	17.3	18.3	19.2	20.7
8.5	13.6	14.1	14.9	16.1	17.6	18.7	19.6	21.2
9.0	13.7	14.2	15.1	16.3	18.0	19.1	20.1	21.8
9.5	13.9	14.4	15.3	16.6	18.3	19.5	20.5	22.4
10.0	14.0	14.5	15.5	16.9	18.7	20.0	21.0	23.0
10.5	14.2	14.7	15.7	17.2	19.1	20.4	21.5	23.6
11.0	14.4	14.9	16.0	17.5	19.5	20.9	22.0	24.1
11.5	14.6	15.2	16.3	17.8	19.9	21.3	22.5	24.7
12.0	14.8	15.4	16.5	18.1	20.2	21.7	23.0	25.3
12.5	15.1	15.7	16.8	18.4	20.6	22.2	23.4	25.8
13.0	15.3	15.9	17.1	18.7	21.0	22.6	23.9	26.3
13.5	15.6	16.2	17.4	19.0	21.3	23.0	24.3	26.8
14.0	15.8	16.4	17.6	19.4	21.7	23.3.	24.7	27.3
14.5	16.1	16.7	17.9	19.6	22.0	23.7	25.1	27.7
15.0	16.3	16.9	18.2	19.9	22.3	24.0	25.5	28.1
15.5	16.6	17.2	18.4	20.2	22.6	24.4	25.8	28.5
16.0	16.8	17.4	18.7	20.5	22.9	24.7	26.1	28.9
16.5	17.0	17.6	18.9	20.7	23.2	24.9	26.4	29.3
17.0	17.2	17.8	19.1	20.9	23.4'	25.2	26.7	29.6
17.5	17.4	18.0	19.3	21.1	23.6	25.4	27.0	30.0
18.0	17.6	18.2	19.5	21.3	23.8	25.7	27.3	30.3
18.5	17.7	18.3	19.6	21.4	24.0	25.9	27.5	30.7
19.0	17.8	18.4	19.7	21.6	24.2	26.1	27.8	31.0
19.5	17.8	18.5	19.8	21.7	24.3	26.3	28.0	31.4
20.0	17.8	18.5	19 8	21.7	24 4	26.5	28.2	31.8

Data used in creating the 2000 Centers for Disease Control (CDC) growth charts, are available at *www.cdc.gov/nchs*, along with technical reports.

Table 6: Length/height (cm), percentiles, with ±3SD values for boys (birth–66 months of age)

Indian affluent

Age in months	N	Percentiles																-3SD	-2SD	-1SD	Median	+1SD	+2SD	+3SD
		3rd	5th	10th	20th	25th	30th	40th	50th	60th	70th	75th	80th	90th	95th	97th							
0	329	47.6	47.8	48.3	49.0	49.1	49.2	49.6	50.1	50.3	50.5	50.8	51.0	51.9	52.4	52.9	46.1	47.5	48.8	50.1	51.5	52.8	54.1
3	386	56.3	56.7	57.3	58.3	58.7	59.1	59.5	60.1	60.6	61.2	61.5	61.7	62.7	63.4	64.3	53.8	55.9	58.0	60.1	62.2	64.3	66.4
6	398	62.0	62.3	63.0	63.7	64.0	64.3	65.0	65.5	66.0	66.6	67.0	67.3	68.9	70.5	71.8	57.9	60.4	63.0	65.5	68.1	70.6	73.1
9	386	67.1	67.4	68.1	69.0	69.3	69.5	69.9	70.3	70.7	71.2	71.6	72.0	73.0	73.8	74.2	64.6	66.5	68.4	70.3	72.2	74.0	75.9
12	373	71.7	72.2	72.7	73.4	73.7	74.0	74.4	74.7	75.1	75.6	75.9	76.1	76.9	78.0	78.6	69.6	71.3	73.0	74.7	76.4	78.1	79.8
18	136	75.3	76.0	76.9	78.5	79.1	79.7	80.2	80.9	81.4	81.9	82.0	82.1	83.3	84.6	85.2	73.6	76.0	78.4	80.9	83.3	85.8	88.2
24	202	78.9	79.9	81.4	82.8	83.4	83.9	85.0	85.7	86.0	86.6	87.1	87.5	88.7	90.1	90.5	76.8	79.8	82.7	85.7	88.6	91.6	94.6
30	266	84.7	85.0	85.8	87.6	88.1	88.6	89.4	90.4	91.0	91.9	92.3	92.7	94.0	95.3	95.9	81.2	84.3	87.4	90.4	93.5	96.5	99.6
36	363	87.7	88.6	89.9	91.8	92.4	92.9	93.9	94.2	94.9	95.9	96.4	96.9	98.5	99.9	100.7	84.7	87.5	90.9	94.2	97.6	101.0	104.3
42	470	89.8	91.1	92.4	94.4	95.1	95.8	96.9	97.9	98.2	99.0	99.9	100.8	102.7	104.1	105.5	85.1	89.8	93.9	97.9	102.0	106.0	110.0
48	523	92.5	94.1	95.4	97.3	97.9	98.5	99.9	100.7	101.9	102.6	103.3	103.9	106.2	108.4	109.4	87.9	92.2	94.4	100.7	105.0	109.2	113.5
54	523	95.6	96.3	98.1	99.9	100.6	101.3	102.5	103.7	104.6	106.0	106.9	107.7	109.9	112.0	113.0	97.9	99.0	100.9	103.7	105.6	107.6	109.5
60	450	97.9	99.3	101.4	103.2	103.9	104.6	105.8	106.9	107.9	109.0	109.9	110.7	113.4	114.9	116.2	92.9	97.6	102.3	106.9	111.6	116.3	121.0
66	277	100.8	102.0	104.3	106.7	107.6	108.4	109.4	110.0	111.0	112.4	113.2	113.9	116.8	118.0	119.4	95.2	100.1	105.0	110.0	114.9	119.8	124.7

N: Number of children   Indian Pediatr 1994; 31:377

Table 7: Weight (kg), percentiles, with ±3SD values for boys (birth–66 months of age)

Indian affluent

Age in months	N	Percentiles																–3SD	–2SD	–1SD	Median	+1SD	+2SD	+3SD
		3rd	5th	10th	20th	25th	30th	40th	50th	60th	70th	75th	80th	90th	95th	97th							
0	329	2.6	2.6	2.7	2.9	3.0	3.0	3.0	3.1	3.2	3.3	3.5	3.4	3.6	3.8	3.9	2.2	2.5	2.8	3.1	3.4	3.7	4.0
3	386	4.7	4.8	5.0	5.3	5.4	5.5	5.6	5.8	5.9	6.1	6.2	6.3	6.7	6.9	7.0	3.9	4.5	5.2	5.3	6.4	7.1	7.7
6	398	6.0	6.1	6.2	6.4	6.5	6.6	6.9	7.1	7.3	7.5	7.7	7.9	8.2	8.5	8.7	4.8	5.5	6.3	7.1	7.8	8.6	9.4
9	386	7.1	7.3	7.5	7.9	8.0	8.1	8.2	8.4	8.5	8.7	8.8	8.9	9.4	9.8	10.0	6.2	6.9	7.6	8.4	9.4	9.1	11.5
12	373	8.2	8.4	8.7	9.0	9.1	9.2	9.5	9.6	9.7	9.9	10.0	10.0	10.4	10.8	11.1	7.5	8.2	8.9	9.6	10.3	11.0	11.7
18	136	8.7	9.0	9.8	10.0	10.1	10.2	10.6	10.9	11.0	11.2	11.4	11.5	12.0	12.5	12.8	7.8	8.9	9.9	10.9	11.9	13.0	14.0
24	202	9.7	10.0	10.4	10.8	11.1	11.2	11.7	11.9	12.0	12.4	12.6	12.8	13.4	14.1	14.9	8.3	9.5	10.9	11.9	13.2	14.4	15.6
30	266	10.7	11.0	11.4	12.0	12.1	12.2	12.5	12.9	13.0	13.5	13.7	13.9	14.7	15.4	15.9	9.0	10.3	11.6	12.9	14.2	15.4	16.7
36	363	11.4	11.5	12.0	12.6	12.8	13.0	13.4	13.7	14.0	14.3	14.6	14.9	15.9	16.5	17.0	9.3	10.7	12.2	13.7	15.2	16.7	18.1
42	470	11.8	12.2	12.7	13.3	13.6	13.8	14.1	14.5	14.9	15.2	15.5	15.8	16.9	17.8	18.5	9.3	11.0	12.8	14.5	16.2	17.9	19.5
48	523	12.4	12.8	13.4	14.0	14.4	14.7	15.0	15.4	15.7	16.1	16.4	16.7	17.7	18.8	19.6	9.9	11.7	13.5	15.4	17.2	19.0	20.8
54	523	13.1	13.7	14.1	14.9	15.0	15.1	15.9	16.2	16.8	17.2	17.6	18.0	19.0	20.0	20.1	10.4	12.3	14.2	16.2	17.8	20.0	21.9
60	450	13.8	14.5	15.0	16.0	16.1	16.2	16.7	17.2	17.8	18.0	18.5	19.0	20.0	20.5	21.8	11.2	13.1	15.1	17.2	19.2	21.3	23.3
66	277	14.4	15.2	16.0	16.9	17.1	17.2	17.7	18.1	18.8	19.4	19.7	20.0	20.9	21.0	22.0	12.1	14.1	17.0	18.1	20.1	22.1	24.1

N: Number of children   Indian Pediatr 1994; 31:377

Table 8: Length/height (cm), percentiles, with ±3SD values for girls (birth–66 months of age)

| | | Percentiles | | | | | | | | | | | | | | | | Indian affluent | | | | |
Age in months	N	3rd	5th	10th	20th	25th	30th	40th	50th	60th	70th	75th	80th	90th	95th	97th	−3SD	−2SD	−1SD	Median	+1SD	+2SD	+3SD
0	269	47.5	48.0	48.3	48.8	49.1	49.3	49.6	50.0	50.3	50.7	51.0	51.2	51.8	52.3	52.7	45.9	47.2	48.6	50.5	51.3	52.7	54.1
3	299	55.3	56.4	57.5	58.3	58.5	58.9	59.3	59.4	60.2	60.5	60.9	61.2	62.0	63.1	63.8	54.1	56.0	57.9	59.3	61.7	63.3	65.5
6	308	61.0	61.7	62.5	63.4	63.7	64.0	64.6	65.2	65.7	66.2	66.6	66.9	68.3	69.0	69.8	58.1	60.5	62.8	65.2	67.5	69.9	72.3
9	302	65.3	66.2	67.3	68.2	68.5	68.7	69.2	69.6	70.1	70.6	70.9	71.1	72.4	73.1	73.9	63.2	65.3	67.5	69.6	71.8	73.9	76.1
12	290	70.2	70.9	72.0	72.7	73.9	73.2	73.6	74.0	74.5	75.0	75.4	75.7	76.5	77.1	77.8	68.5	70.4	72.2	74.0	75.9	77.0	79.4
18	135	73.5	74.8	75.9	78.0	78.5	78.9	79.7	80.0	80.5	81.0	81.5	82.0	83.4	84.7	85.0	73.7	75.8	77.9	80.0	82.1	84.2	86.3
24	179	77.7	79.0	80.6	82.0	82.7	83.4	84.0	84.4	85.0	85.9	86.2	86.5	88.1	89.4	89.8	76.0	79.0	82.0	84.9	88.0	90.0	93.8
30	206	82.0	82.7	84.7	86.4	86.9	87.4	88.6	89.3	90.0	90.7	91.3	91.8	93.0	94.2	95.0	78.3	82.0	85.7	89.3	93.0	96.7	100.4
36	266	85.3	86.1	87.9	89.9	90.7	91.4	92.4	92.9	94.0	94.9	95.4	95.8	97.5	99.1	99.9	81.5	85.3	89.1	92.9	96.8	100.6	104.4
42	370	88.1	89.9	91.0	93.2	93.0	94.4	95.6	96.7	97.4	98.3	108.7	99.0	101.4	102.7	104.5	85.2	89.0	92.0	96.7	100.5	104.4	108.2
48	432	91.2	92.9	94.9	96.9	97.4	97.9	98.7	99.9	100.7	101.8	102.4	102.9	104.8	106.9	107.9	88.4	91.6	95.8	99.9	104.1	109.3	112.5
54	459	94.5	95.8	97.3	99.9	100.5	101.0	102.0	103.0	104.0	105.0	105.5	106.0	107.0	109.0	110.0	90.7	94.8	98.9	103.0	107.0	111.2	115.3
60	381	96.9	98.4	100.1	102.4	103.1	103.7	104.9	106.0	107.3	108.5	109.1	109.7	111.1	112.5	113.7	92.7	97.1	101.5	106.0	110.4	114.8	119.3
66	245	100.8	102.0	103.6	105.6	106.3	107.0	108.1	109.4	110.6	111.6	112.2	112.8	114.7	116.0	117.0	96.0	100.5	104.9	109.4	113.8	118.3	122.7

N: Number of children

Indian Pediatr 1994; 31:377

Table 9: Weight (kg), percentiles, with ±3SD values for girls (birth–66 months of age)

Age in months	N	Percentiles															Indian affluent						
		3rd	5th	10th	20th	25th	30th	40th	50th	60th	70th	75th	80th	90th	95th	97th	-3SD	-2SD	-1SD	Median	+1SD	+2SD	+3SD
0	269	2.6	2.6	2.7	2.8	2.9	2.9	3.0	3.1	3.2	3.3	3.4	3.5	3.7	3.8	3.9	1.9	2.3	2.7	3.1	3.5	3.9	4.2
3	299	4.4	4.5	4.8	5.1	5.2	5.3	5.4	5.6	5.7	5.9	6.0	6.1	6.4	6.6	6.7	3.3	4.4	5.0	5.6	6.2	6.8	7.4
6	308	5.6	5.7	5.9	6.2	6.4	6.5	6.7	6.9	7.1	7.2	7.4	7.5	7.9	8.2	8.4	4.7	5.4	6.2	6.9	7.6	8.4	9.1
9	302	6.9	7.0	7.3	7.4	7.5	7.6	7.8	8.0	8.2	8.4	8.5	8.6	8.9	9.2	9.5	5.9	6.6	7.3	8.0	8.7	9.4	10.1
12	290	7.8	8.0	8.2	8.5	8.6	8.7	9.0	9.1	9.3	9.5	9.7	9.8	10.0	10.4	10.6	7.0	7.7	8.4	9.1	9.8	10.5	11.2
18	135	8.7	9.0	9.5	9.9	10.0	10.1	10.3	10.5	10.8	10.9	11.4	11.2	11.8	12.1	12.4	7.8	8.7	9.6	10.5	11.4	12.3	13.3
24	179	9.4	9.6	10.0	10.7	10.9	11.0	11.2	11.6	11.8	12.0	12.2	12.4	13.0	13.5	13.9	8.1	9.3	10.4	11.5	12.7	13.9	15.0
30	206	9.9	10.2	10.6	11.4	11.7	12.0	12.2	12.5	12.8	12.9	13.0	13.4	13.9	14.4	14.8	8.7	10.0	11.2	12.5	13.7	15.0	16.2
36	266	10.5	10.8	11.4	12.1	12.4	12.6	13.0	13.4	13.7	14.0	14.2	14.3	15.0	15.7	16.4	9.0	10.4	12.0	13.4	14.9	16.4	18.0
42	370	11.4	11.8	12.1	12.9	13.2	13.5	13.9	14.2	14.6	14.9	15.2	15.4	16.5	17.2	17.8	9.2	10.9	12.5	14.2	15.1	16.8	18.4
48	432	11.9	12.2	12.9	13.6	13.9	14.1	14.6	15.0	15.5	16.0	16.5	16.9	17.5	18.2	18.9	9.5	11.3	13.2	15.0	16.9	18.7	20.5
54	459	12.9	13.3	14.0	14.5	14.8	15.0	15.4	16.0	16.7	17.0	17.5	17.9	18.7	19.7	20.0	1.2	12.1	14.1	16.0	17.9	19.8	21.8
60	381	13.4	13.9	14.5	15.0	15.4	15.7	16.2	17.0	17.4	17.9	18.4	18.8	19.9	20.0	21.0	10.6	12.7	14.8	17.0	19.0	21.2	23.3
66	245	14.2	14.9	15.6	15.6	16.2	16.6	17.4	17.9	18.4	18.9	19.2	19.4	20.1	21.0	21.4	11.9	14.0	16.3	17.9	20.5	22.6	24.7

N: Number of children   Indian Pediatr 1994; 31:377

Table 10: *Means for height (cm), SD± and percentiles for boys*

(6–18 years of age)   *Indian affluent*

Age (yrs)	N	Mean	SD±	Percentiles							
				3rd	5th	10th	25th	50th	75th	90th	97th
6.0	175	113.7	5.4	103.7	105.5	106.9	112.0	114.2	118.0	122.6	125.9
6.5	128	117.5	5.5	106.1	107.5	109.6	113.5	117.3	120.7	127.2	128.4
7.0	235	118.6	5.5	108.5	109.8	112.0	115.9	119.7	123.0	126.9	130.8
7.5	213	121.6	5.8	110.9	111.3	114.2	117.8	121.6	125.2	130.0	133.2
8.0	295	124.1	5.5	113.3	114.4	116.3	119.7	123.6	127.4	132.4	135.8
8.5	275	126.4	6.1	115.2	116.2	118.6	121.9	125.7	129.8	137.0	138.5
9.0	338	130.4	6.3	118.0	118.5	120.9	124.2	128.2	132.5	138.7	141.4
9.5	348	131.5	6.5	120.3	120.9	123.4	126.7	130.8	135.3	142.1	144.5
10.0	425	134.7	6.4	122.7	123.4	125.9	129.4	133.6	138.3	144.0	147.7
10.5	612	137.6	6.4	125.1	125.9	128.6	132.2	136.6	141.5	148.0	151.0
11.0	621	139.6	7.0	127.5	128.5	131.2	135.6	139.6	144.7	150.1	154.3
11.5	761	142.3	7.2	129.9	131.1	133.9	138.0	142.7	147.9	152.3	157.5
12.0	755	144.7	7.7	132.4	133.8	136.6	141.0	145.8	151.1	157.0	160.8
12.5	889	147.9	8.1	134.9	136.5	139.3	143.9	148.9	154.2	160.5	163.9
13.0	771	150.3	8.1	137.4	139.2	142.0	146.8	152.0	157.3	163.0	166.9
13.5	829	154.9	8.1	140.0	141.8	144.7	149.7	154.9	160.2	167.5	169.7
14.0	754	158.0	8.0	142.6	144.5	147.4	152.4	157.6	162.9	170.1	172.5
14.5	743	161.4	7.4	145.2	147.2	149.9	155.0	160.2	165.4	172.0	174.7
15.0	628	164.3	6.8	148.0	149.8	152.4	157.4	162.5	167.7	173.0	176.8
15.5	528	165.5	6.4	150.8	152.4	154.9	159.6	164.6	169.6	174.0	178.5
16.0	461	167.1	6.2	153.6	154.9	157.9	161.6	166.3	171.2	176.2	179.8
16.5	393	167.9	6.3	156.6	157.4	159.4	163.3	167.7	172.4	176.8	180.7
17.0	288	168.6	6.1	159.6	159.8	161.5	165.0	168.7	173.1	176.0	181.2
17.5	177	169.4	5.6	162.7	163.1	163.4	167.5	169.3	173.4	177.0	181.1
18.0	87	168.9	5.6	161.0	163.5	164.0	168.8	169.8	173.4	177.1	181.6

N: Number of subjects

SD: Standard deviation

Indian Pediatr 1992; 29:1203

Table 11: *Means for weight (kg), SD± and percentiles for boys*

(6–18 years of age)

Indian affluent

Age (yrs)	N	Mean	SD±	Percentiles							
				3rd	5th	10th	25th	50th	75th	90th	97th
6.0	175	19.2	0.9	15.2	15.7	16.2	18.0	19.0	20.7	24.7	25.4
6.5	128	20.6	1.1	15.7	16.4	17.4	18.6	20.0	21.9	24.2	27.7
7.0	235	21.0	1.0	16.2	16.9	18.2	19.4	21.0	22.9	25.7	29.7
7.5	213	22.4	1.1	16.8	17.5	18.7	20.0	22.0	23.9	28.1	31.6
8.0	295	23.5	1.1	17.5	18.0	19.1	20.7	22.6	25.0	29.6	33.5
8.5	275	24.5	1.3	18.2	18.6	19.7	21.3	23.5	26.3	31.6	35.5
9.0	338	26.5	1.3	19.2	19.4	20.3	22.0	24.4	27.7	32.7	37.7
9.5	348	26.8	1.3	19.9	20.2	21.2	22.9	25.6	29.4	35.0	40.1
10.0	425	28.7	1.4	20.9	21.2	22.3	24.1	27.0	31.3	37.7	42.7
10.5	612	30.8	1.4	21.9	22.3	23.5	25.5	28.7	33.4	40.7	45.4
11.0	621	31.9	1.5	22.9	23.5	24.9	27.1	30.6	35.6	42.2	48.2
11.5	761	33.8	1.6	24.1	24.9	26.3	28.9	32.7	37.9	45.6	51.1
12.0	755	35.4	1.5	25.3	26.3	27.9	30.7	34.8	40.3	47.8	54.1
12.5	889	37.9	1.7	26.7	27.8	29.6	32.7	37.1	42.7	49.4	57.1
13.0	771	39.4	2.0	28.1	29.3	31.3	34.7	39.4	45.1	53.8	60.0
13.5	829	43.2	2.2	29.6	31.0	33.1	36.8	41.8	47.6	56.9	63.0
14.0	754	44.7	2.3	31.2	32.7	34.9	38.8	44.1	50.0	58.3	65.9
14.5	743	48.1	2.4	32.9	34.5	36.7	40.9	46.3	52.4	62.2	68.7
15.0	628	51.0	2.6	34.6	36.3	38.6	42.8	48.5	54.6	65.1	71.4
15.5	528	52.4	3.1	36.5	38.1	40.3	44.7	50.5	56.8	69.0	73.9
16.0	461	55.0	3.6	38.5	40.0	42.1	46.5	52.4	58.8	69.7	76.3
16.5	393	54.9	3.5	40.6	41.9	43.8	48.1	54.0	60.6	68.3	78.5
17.0	288	56.6	4.1	42.8	43.9	45.9	50.0	55.5	62.3	67.2	80.5
17.5	177	56.9	5.0	45.2	45.8	46.9	52.0	57.2	63.7	72.3	82.2
18.0	87	59.7	9.3	47.6	47.8	48.3	54.0	58.6	64.9	80.0	83.6

N: Number of subjects
SE: Standard deviations

Indian Pediatr 1992; 29:1203

Table 12: *Means for height (cm), SD± and percentiles for girls*

(6–17 years of age)   *Indian affluent*

Age (yrs)	N	Mean	SD±	Percentiles							
				3rd	5th	10th	25th	50th	75th	90th	97th
6.0	241	113.0	4.7	102.1	104.5	106.1	108.8	112.5	115.9	120.5	123.3
6.5	251	115.4	5.1	104.5	107.0	108.4	111.1	114.9	118.4	124.0	126.0
7.0	294	118.2	6.0	107.1	109.4	110.7	113.7	117.4	121.3	126.5	129.3
7.5	319	120.2	5.5	109.7	111.6	113.1	116.4	120.3	124.4	128.9	132.8
8.0	328	122.7	5.8	112.3	113.9	115.5	119.3	123.2	127.5	133.2	136.4
8.5	349	126.2	6.2	115.0	116.2	118.1	122.2	126.2	130.7	136.6	139.8
9.0	399	128.6	6.4	117.8	118.8	120.9	125.1	129.2	133.8	138.3	143.1
9.5	429	131.9	6.6	120.6	121.4	123.6	128.0	132.3	136.9	142.1	146.2
10.0	487	134.8	6.8	123.4	124.1	126.5	130.8	135.2	139.8	145.0	149.0
10.5	452	137.9	7.2	126.1	126.9	129.3	133.7	138.1	142.7	148.6	151.7
11.0	503	141.3	2.3	128.8	129.7	132.1	136.4	140.9	145.4	152.1	154.2
11.5	490	144.3	6.9	131.4	132.4	134.8	139.0	143.5	147.9	154.7	156.5
12.0	435	146.7	6.7	133.9	135.0	137.4	141.5	146.0	150.3	156.5	158.5
12.5	489	149.9	6.2	136.3	137.5	139.8	143.8	148.3	152.5	158.5	160.4
13.0	455	151.4	6.0	138.5	139.8	142.1	145.9	150.4	154.4	161.0	162.1
13.5	456	153.2	6.0	140.6	141.9	144.1	147.8	152.2	156.2	162.5	163.5
14.0	391	153.6	5.7	142.4	143.8	145.9	149.4	153.8	157.6	162.0	164.7
14.5	350	154.8	5.6	144.1	145.4	147.4	150.8	155.1	158.8	161.3	165.8
15.0	291	155.0	5.6	145.5	146.6	148.6	151.8	156.0	159.7	163.5	166.5
15.5	204	155.4	5.6	146.6	147.5	149.3	152.6	156.6	160.4	163.5	167.1
16.0	176	155.1	5.0	147.5	148.0	149.7	152.9	156.8	160.4	162.5	167.4
16.5	182	156.0	5.1	148.0	148.1	149.7	152.9	156.5	160.5	163.0	167.6
17.0	116	157.1	5.9	148.3	148.5	149.8	153.0	157.0	160.5	165.0	168.0

N: Number of subjects   Indian Pediatr 1992; 29:1203
SD: Standard deviation

Table 13: *Means for weight (kg), SD± and percentiles for girls*

(6–17 years of age)   *Indian affluent*

Age (yrs)	N	Mean	SD±	Percentiles							
				3rd	5th	10th	25th	50th	75th	90th	97th
6.0	241	18.7	2.9	14.1	15.2	15.7	16.4	17.8	19.2	22.9	23.7
6.5	251	19.6	3.2	14.4	15.5	16.1	16.9	18.3	19.9	23.9	25.4
7.0	294	20.5	3.4	14.8	15.8	16.4	17.3	19.0	20.9	26.1	27.5
7.5	319	21.7	4.3	15.3	16.2	16.9	18.0	19.9	22.2	27.8	29.8
8.0	328	23.0	4.3	15.9	16.4	17.2	18.7	20.8	23.6	31.9	32.3
8.5	349	24.9	5.2	16.4	16.8	17.8	19.6	22.0	25.3	33.0	34.9
9.0	399	25.8	5.2	17.1	17.6	18.7	20.7	23.5	27.2	35.3	37.7
9.5	429	27.5	6.0	18.3	18.5	19.7	22.1	25.1	29.3	37.0	40.5
10.0	487	29.6	6.8	19.5	19.7	21.0	23.6	26.9	31.4	40.0	43.4
10.5	452	31.9	7.2	20.9	21.0	22.4	25.3	28.9	33.7	45.4	46.4
11.0	503	34.3	8.1	22.3	22.4	24.0	27.1	30.9	36.0	47.5	49.3
11.5	490	36.8	8.2	23.7	24.0	25.6	28.9	32.9	38.4	49.6	52.2
12.0	435	38.7	8.6	25.1	25.6	27.3	30.8	35.0	40.7	53.0	55.1
12.5	489	41.9	8.8	26.5	27.2	29.0	32.6	37.1	42.9	54.9	57.9
13.0	455	42.6	8.5	27.9	28.9	30.7	34.5	39.1	45.1	57.0	60.7
13.5	456	45.2	9.4	29.3	30.6	32.4	36.2	41.0	47.1	57.2	63.2
14.0	391	45.7	8.9	30.7	32.1	34.0	37.8	42.7	48.9	57.7	65.7
14.5	350	46.6	8.4	32.1	33.6	35.5	39.3	44.3	50.5	58.4	67.9
15.0	291	48.0	9.2	33.4	35.0	36.9	40.6	45.7	51.8	61.2	70.0
15.5	204	48.9	9.1	34.6	36.2	38.2	41.7	46.8	52.9	61.6	71.8
16.0	176	49.2	9.2	35.7	37.3	39.3	42.5	47.7	53.6	63.4	73.3
16.5	182	49.6	8.4	36.7	38.1	40.1	43.0	48.2	54.0	58.9	74.6
17.0	116	49.0	7.8	37.6	38.7	40.7	43.3	48.4	53.9	59.8	75.6

N: Number of subjects   Indian Pediatr 1992; 29:1203
SD: Standard deviation

Table 14: Head circumference (cm), percentiles, with ±3SD values for boys (birth–6 years of age)

Indian affluent

Age in months	N	Percentiles																–3SD	–2SD	–1SD	Median	+1SD	+2SD	+3SD
		3rd	5th	10th	20th	25th	30th	40th	50th	60th	70th	75th	80th	90th	95th	97th							
0	329	33.2	33.5	33.8	34.1	34.2	34.3	34.4	34.7	34.9	35.1	35.3	35.4	35.9	36.3	36.5	32.1	33.0	33.3	34.7	35.5	36.3	37.1
3	386	38.1	38.2	38.6	39.1	39.3	39.5	39.7	40.0	40.2	40.5	40.7	40.9	41.4	42.0	42.5	36.5	37.6	38.8	40.0	41.1	42.3	43.5
6	398	40.3	40.7	41.3	41.9	42.1	42.2	42.5	42.7	42.9	43.1	43.3	43.4	43.8	44.2	44.7	39.6	40.6	41.7	42.7	43.7	44.8	45.8
9	386	42.5	42.9	43.2	43.6	43.7	43.8	44.0	44.2	44.4	44.6	44.8	44.9	45.3	45.8	46.2	41.6	42.4	43.3	44.2	45.1	46.0	46.9
12	373	43.7	44.0	44.4	44.8	44.9	45.0	45.2	45.4	45.5	45.8	45.9	46.0	46.5	46.9	47.3	42.2	43.7	44.5	54.4	46.2	47.1	48.0
18	136	44.8	44.9	45.4	46.0	46.1	46.2	46.7	47.0	47.5	47.9	48.0	48.5	48.4	48.7	48.9	43.4	44.6	45.8	47.0	48.2	49.4	50.6
24	202	45.5	45.8	46.3	46.8	46.9	47.0	47.3	47.7	48.0	48.4	48.5	48.5	49.0	49.5	49.9	44.3	45.4	46.6	47.7	48.8	50.6	51.2
30	266	46.2	46.4	46.8	47.3	47.4	47.5	47.9	48.2	48.5	49.4	49.5	48.9	49.6	49.9	50.4	44.8	46.0	47.0	48.2	49.3	50.4	51.3
36	363	46.1	46.7	47.0	47.8	47.9	48.0	48.4	48.7	49.0	49.9	50.1	49.7	50.0	50.9	51.0	44.8	46.1	47.4	48.7	50.6	51.3	52.6
42	470	46.9	47.1	47.5	48.0	48.2	48.4	48.9	49.3	49.6	49.9	50.1	50.3	50.9	51.0	51.4	45.4	46.8	48.0	49.3	50.6	51.8	53.2
48	523	46.9	47.1	47.9	48.5	48.7	48.9	49.4	49.9	50.0	50.5	50.7	50.9	51.2	52.0	52.1	45.6	47.0	48.5	49.9	51.3	52.7	54.1
54	523	47.4	47.7	48.0	48.8	48.9	49.2	49.6	50.0	50.4	50.9	51.0	51.2	52.0	52.5	52.9	55.5	57.0	58.5	60.0	61.4	62.9	64.4
60	450	47.5	47.9	48.4	48.9	49.2	49.5	50.0	50.3	50.6	51.0	51.2	51.5	52.3	52.9	53.0	45.6	47.1	48.7	50.3	51.8	53.4	55.5
66	277	47.7	48.3	48.8	49.4	49.6	49.9	50.4	50.9	51.3	51.6	51.9	52.1	53.0	53.5	53.9	46.1	47.6	49.3	50.9	52.5	54.1	55.7
72	124	47.9	48.3	48.8	49.5	49.8	50.1	50.8	51.1	51.6	52.0	52.2	52.4	53.3	54.2	54.5	45.9	47.6	49.4	51.1	52.8	54.6	56.3

Indian Pediatr 1994; 31:377

N: Number of children

Table 15: Head circumference (cm), percentiles, with ±3SD values for girls (birth–6 years of age)

Indian affluent

Age in months	N	3rd	5th	10th	20th	25th	30th	40th	50th	60th	70th	75th	80th	90th	95th	97th	-3SD	-2SD	-1SD	Median	+1SD	+2SD	+3SD
0	269	33.1	33.2	33.6	33.9	34.1	34.2	34.4	34.5	34.3	35.0	35.2	35.3	35.5	36.0	36.4	32.1	32.9	33.8	34.6	35.5	36.3	37.2
3	299	37.4	37.3	38.2	38.8	39.0	39.1	39.4	39.6	39.9	40.1	40.3	40.5	41.0	41.3	41.4	36.4	37.5	38.6	39.6	40.7	41.8	42.8
6	308	39.7	40.1	40.3	41.6	41.8	42.0	42.2	42.4	42.6	42.8	42.9	43.0	43.4	43.8	44.0	39.3	40.3	41.4	42.4	43.5	44.5	45.6
9	302	42.2	42.5	42.8	43.2	43.4	43.5	43.7	43.9	44.1	44.3	44.5	44.6	44.9	45.3	45.5	41.4	42.2	43.1	43.9	44.7	45.6	46.4
12	290	43.1	43.5	44.1	44.4	44.6	44.7	44.9	45.1	45.3	45.5	45.7	45.8	46.1	46.4	46.5	42.4	43.3	44.2	45.1	46.0	47.0	47.9
18	135	43.9	44.4	44.9	45.5	45.7	45.9	46.0	46.4	46.7	46.9	47.0	47.2	47.8	48.0	48.0	42.7	44.6	45.2	46.4	47.6	48.9	50.1
24	179	44.7	45.0	45.5	46.0	46.2	46.5	46.9	47.0	47.3	47.5	47.7	47.9	48.3	48.7	48.9	43.5	44.6	45.8	47.0	48.1	49.3	50.5
30	206	44.9	45.4	45.8	46.5	46.7	47.0	47.5	47.8	48.0	48.2	48.4	48.5	49.1	49.7	50.0	43.3	45.1	46.5	47.8	49.1	50.5	51.8
36	266	45.3	45.7	46.0	46.8	47.0	47.3	47.6	48.0	48.5	48.9	49.0	49.1	49.9	50.4	50.9	43.9	45.1	46.6	47.9	49.9	50.8	52.3
42	370	45.9	46.0	46.5	47.2	47.5	47.7	48.0	48.4	48.9	49.4	49.7	49.9	50.5	51.0	51.5	43.8	45.4	46.9	48.4	50.0	51.5	53.0
48	432	45.9	46.4	46.9	47.5	47.7	47.9	48.4	48.9	49.3	49.9	50.2	50.4	51.0	51.8	52.0	43.5	45.3	47.1	48.9	50.7	52.5	54.3
54	459	46.2	46.9	47.2	47.9	48.1	48.3	48.8	49.2	49.5	50.0	50.4	50.8	51.5	52.0	52.5	44.2	45.9	47.5	49.2	50.9	52.6	54.2
60	381	46.6	47.2	47.5	48.4	48.7	48.9	49.1	49.5	49.9	50.4	50.7	51.0	51.9	52.6	53.0	44.5	46.2	47.9	49.2	51.2	52.9	54.5
66	245	47.4	47.5	48.0	48.7	48.9	49.1	49.5	50.0	50.5	51.0	51.3	51.5	52.1	53.0	53.3	45.0	46.6	48.3	50.0	51.5	53.3	55.0
72	98	47.5	47.7	48.1	48.9	49.2	49.6	50.4	51.3	51.8	52.0	52.2	52.4	53.1	54.2	54.9	45.3	47.3	49.3	51.3	53.3	55.3	57.3

N: Number of children

Indian Pediatr 1994; 31:377

Table 16: Mid-arm circumference (cm), with ±3SD values for boys (birth–6 years of age)

Indian affluent

Age in months	N	3rd	5th	10th	20th	25th	30th	40th	50th	60th	70th	75th	80th	90th	95th	97th	−3SD	−2SD	−1SD	Median	+1SD	+2SD	+3SD
							Percentiles																
0	329	9.0	9.0	9.1	9.4	9.5	9.6	9.3	9.9	10.0	10.2	10.3	10.4	10.7	11.1	11.2	8.1	8.7	9.4	9.9	10.5	11.1	11.3
3	386	9.7	9.9	10.2	10.7	11.0	11.2	11.7	12.1	12.4	12.7	12.9	13.0	13.3	14.1	14.4	8.4	9.6	10.9	12.1	13.4	14.5	15.5
6	398	10.5	10.7	11.1	11.5	11.9	12.2	13.0	13.6	13.8	14.0	14.2	14.4	14.9	15.4	15.8	9.1	10.8	12.1	13.6	15.0	16.5	18.0
9	386	11.3	11.5	12.0	12.6	12.9	13.1	13.7	14.4	14.7	15.0	15.2	15.3	15.7	16.0	16.4	10.0	11.1	12.9	14.4	15.8	17.3	18.7
12	373	12.5	12.7	13.2	13.7	13.9	14.0	14.5	14.9	15.2	15.4	15.6	15.7	16.0	16.3	16.5	11.4	12.6	13.8	14.9	16.1	17.2	18.4
18	136	12.8	13.6	13.7	14.0	14.1	14.2	14.4	14.6	14.9	15.0	15.4	15.7	16.0	16.5	16.8	11.5	12.5	13.5	14.5	15.6	16.5	17.6
24	202	13.0	13.5	13.9	14.2	14.3	14.5	14.6	14.9	15.0	15.3	15.7	16.0	16.5	16.8	17.1	11.8	12.8	13.9	14.9	16.0	17.0	18.1
30	266	13.5	13.3	14.4	14.5	14.7	14.9	15.0	15.4	15.8	16.0	16.2	16.5	16.9	17.3	17.6	12.1	13.2	14.3	15.4	16.4	17.5	18.6
36	363	13.7	13.9	14.3	14.7	14.8	14.9	15.0	15.4	15.7	16.0	16.2	16.5	17.0	17.6	17.9	12.0	13.1	14.2	15.4	16.5	17.6	18.8
42	470	13.3	14.1	14.4	14.9	14.9	15.0	15.4	15.5	15.9	16.2	16.4	16.5	17.3	18.0	18.3	11.9	13.1	14.3	15.5	16.7	17.8	19.0
48	523	14.2	14.3	14.5	14.9	15.2	15.4	15.6	15.9	16.0	16.5	16.7	17.0	17.9	18.0	18.5	11.7	13.1	14.5	15.9	17.3	18.7	28.1
54	523	14.3	14.5	14.7	15.0	15.2	15.4	15.8	16.0	16.2	16.5	16.7	17.0	17.9	18.2	18.7	9.9	11.9	13.9	16.0	18.0	20.0	22.1
60	450	14.4	14.6	14.9	15.3	15.5	15.6	16.0	16.3	16.5	16.9	17.0	17.1	17.9	18.3	18.9	12.1	13.5	14.9	16.3	17.7	19.1	20.5
66	277	14.3	14.5	15.0	15.4	15.7	15.9	16.2	16.5	16.9	17.1	17.3	17.5	18.0	18.7	19.0	12.6	13.9	15.2	16.5	17.8	19.1	20.4
72	124	14.2	14.4	15.2	15.5	15.8	16.0	16.4	16.7	17.3	17.6	17.8	18.0	18.6	19.6	20.1	13.1	14.3	15.5	16.7	17.9	19.1	20.3

N: Number of children

Indian Pediatr 1994; 31:377

Table 17: Mid-arm circumference (cm) percentiles, with ±3SD values for girls (birth–6 years of age)

Age in months	N	Percentiles															Indian affluent						
		3rd	5th	10th	20th	25th	30th	40th	50th	60th	70th	75th	80th	90th	95th	97th	–3SD	–2SD	–1SD	Med-ian	+1SD	+2SD	+3SD
0	269	8.9	9.0	9.2	9.4	9.6	9.7	9.8	9.9	10.0	10.2	10.4	10.5	10.9	11.2	11.2	8.1	8.7	9.3	9.9	10.5	11.2	11.8
3	299	9.9	10.1	10.4	10.8	11.1	11.3	11.7	12.1	12.3	12.5	12.7	12.8	13.2	13.5	13.6	8.9	9.9	11.0	12.1	13.1	14.2	15.2
6	308	10.5	10.6	11.0	11.5	11.9	12.2	12.3	13.3	13.6	13.9	14.0	14.1	14.5	14.9	15.1	9.3	10.5	12.0	13.3	14.7	16.1	17.4
9	302	11.1	11.3	11.7	12.3	12.6	12.8	13.6	14.2	14.5	14.8	14.9	15.0	15.3	15.5	15.9	10.0	11.4	13.3	14.2	15.6	17.0	18.4
12	290	12.1	12.3	12.6	13.2	13.4	13.6	14.2	14.6	15.0	15.2	15.2	15.2	15.9	16.4	16.5	10.9	12.1	13.4	14.6	15.9	17.1	18.3
18	135	12.3	12.6	13.3	13.9	14.0	14.1	14.3	14.4	14.6	14.9	15.2	15.5	15.8	16.4	16.5	11.3	12.3	13.4	14.4	15.5	16.5	17.5
24	179	13.0	13.3	13.6	14.1	14.2	14.4	14.6	14.9	15.0	15.4	15.6	15.7	16.4	16.5	17.0	11.3	12.8	13.8	14.9	16.0	17.1	18.1
30	206	13.4	13.5	14.1	14.4	14.6	14.8	15.0	15.3	15.6	15.9	16.0	16.2	16.7	17.0	17.2	12.1	13.2	14.3	15.3	16.4	17.4	13.5
36	266	13.6	13.7	14.1	14.6	14.8	15.0	15.2	15.4	15.7	16.0	16.2	16.4	17.0	17.6	18.0	11.9	13.1	14.3	15.4	16.6	17.8	19.0
42	370	13.9	14.2	14.4	14.8	15.0	15.2	15.5	15.7	15.9	16.2	16.4	16.5	17.1	17.8	18.0	12.5	13.5	14.6	15.7	16.7	17.8	18.9
48	432	13.9	14.2	14.5	14.9	15.1	15.3	15.7	15.9	16.2	16.5	16.7	17.0	17.8	18.0	18.5	12.2	13.5	14.7	15.9	17.2	18.4	19.7
54	459	14.0	14.2	14.7	15.1	15.3	15.5	15.8	16.0	16.4	16.8	16.9	17.0	17.7	18.0	18.5	12.5	13.7	14.9	16.0	17.2	18.4	19.6
60	381	13.9	14.2	14.9	15.2	15.4	15.6	15.9	16.2	16.5	16.9	17.0	17.2	17.9	18.2	18.6	12.6	13.8	15.0	16.2	17.4	18.6	19.8
66	245	14.2	14.4	14.8	15.4	15.7	15.9	16.0	16.4	16.6	17.0	17.2	17.4	17.9	18.2	18.8	12.6	13.9	15.1	16.4	17.6	18.9	20.1
72	98	14.3	14.6	15.0	15.7	15.9	16.0	16.3	16.7	17.3	17.6	17.8	18.0	18.3	19.1	19.2	12.6	14.0	15.4	16.7	18.1	19.5	20.9

N: Number of children

Indian Pediatr 1994; 31:377

Table 18: Chest circumference (cm) percentiles, with ±3SD values for girls (birth–6 years of age)

| Age in months | N | Percentiles | | | | | | | | | | | | | | | | Indian affluent | | | | | | |
		3rd	5th	10th	20th	25th	30th	40th	50th	60th	70th	75th	80th	90th	95th	97th	-3SD	-2SD	-1SD	Median	+1SD	+2SD	+3SD
0	329	31.0	31.1	31.3	31.8	32.0	32.1	32.3	32.5	32.9	33.2	33.4	33.5	34.0	34.4	34.7	29.4	30.4	31.4	32.5	33.5	34.6	35.6
3	386	36.1	36.3	36.9	37.5	37.8	38.0	38.3	38.7	39.2	39.5	39.9	40.2	41.0	41.8	42.3	33.9	35.5	37.1	38.7	40.3	42.0	43.6
6	398	39.1	39.3	39.5	40.5	40.8	41.0	41.4	41.7	42.1	42.5	42.8	43.0	44.1	45.5	46.3	36.5	38.2	40.0	41.7	43.5	45.2	47.0
9	386	41.2	41.7	42.2	42.8	43.0	43.2	43.6	43.9	44.3	44.7	44.9	45.1	46.4	48.2	48.9	38.3	40.2	42.0	43.9	45.8	47.6	49.5
12	373	43.0	43.3	43.9	44.5	44.8	45.0	45.3	45.5	45.8	46.1	46.4	46.7	49.3	49.5	50.5	40.3	42.0	43.8	45.5	47.3	49.1	50.8
18	136	44.6	44.9	45.6	46.4	46.7	47.0	47.4	47.8	48.4	48.8	49.1	49.4	50.0	51.0	52.0	42.2	44.0	45.9	47.8	49.7	51.5	53.4
24	202	44.8	45.6	46.8	47.5	47.7	48.0	48.4	48.9	49.2	49.8	50.1	50.4	51.6	52.5	54.0	42.2	44.4	46.7	48.9	51.2	53.4	55.7
30	266	46.2	46.8	47.5	48.5	48.7	49.0	49.4	49.9	50.0	50.9	51.2	51.5	52.8	53.8	54.7	43.3	45.5	47.7	49.9	52.1	54.3	56.5
36	363	46.9	47.8	48.3	49.3	49.6	50.0	50.2	50.9	51.0	51.8	52.1	52.4	53.7	54.5	55.6	44.5	46.6	48.8	50.9	53.0	55.2	57.3
42	470	47.5	48.4	48.9	49.9	50.4	50.9	51.3	51.9	52.1	52.9	53.2	53.4	54.9	56.0	56.8	44.9	47.2	49.6	51.9	54.3	56.6	58.9
48	523	48.0	48.5	49.3	50.3	50.7	51.0	52.0	52.5	53.0	53.5	53.8	54.2	54.4	57.0	57.6	43.5	46.5	49.5	52.5	55.0	58.5	61.5
54	523	48.6	49.0	50.0	50.9	51.3	51.6	52.3	52.9	53.4	54.0	54.4	54.9	56.0	57.8	58.8	43.7	46.8	49.9	52.9	56.0	59.0	62.1
60	450	49.0	49.8	50.6	51.3	51.9	52.4	53.0	53.9	54.3	55.0	55.3	55.7	57.0	57.9	59.0	46.1	48.7	51.3	53.9	56.5	59.1	61.7
66	277	50.6	50.9	51.7	52.3	52.6	53.0	53.9	54.8	55.2	55.8	56.1	56.5	57.7	58.5	59.6	47.2	49.8	52.3	54.8	57.3	59.8	62.4
72	124	51.1	51.7	52.0	53.0	53.5	54.1	54.9	55.8	56.2	57.0	57.4	57.9	59.0	61.4	61.9	47.3	50.2	53.0	55.8	58.7	61.5	64.3

N: Number of children

Indian Pediatr 1994; 31:377

Table 19: Chest circumference (cm) percentiles, with ±3SD values for girls (birth–6 years of age)

Age in months	N	Percentiles															Indian affluent						
		3rd	5th	10th	20th	25th	30th	40th	50th	60th	70th	75th	80th	90th	95th	97th	-3SD	-2SD	-1SD	Median	+1SD	+2SD	+3SD
0	269	30.3	30.5	31.1	31.4	31.6	31.8	32.1	32.3	32.5	32.9	33.2	33.4	33.9	34.3	34.7	29.0	30.1	31.2	32.3	33.4	34.5	35.6
3	299	35.9	36.3	36.8	37.4	37.6	37.8	38.1	38.3	38.6	39.1	39.4	39.6	40.4	41.0	41.5	34.0	35.4	36.9	38.3	39.8	41.2	42.7
6	308	38.5	39.0	39.5	40.1	40.3	40.5	41.0	41.3	41.5	41.9	42.2	42.4	43.4	44.3	45.3	36.4	38.0	39.6	41.3	42.9	44.5	46.1
9	302	40.6	41.1	41.6	42.1	42.4	42.7	43.2	43.6	43.9	44.2	44.5	44.7	45.3	46.3	47.0	38.8	40.4	42.0	43.6	45.2	46.8	48.4
12	290	42.3	43.0	43.5	44.2	44.4	44.6	45.0	45.3	45.6	45.9	46.2	46.4	47.6	48.7	49.3	40.3	41.9	43.6	45.3	47.0	48.7	50.3
18	135	43.5	43.8	45.0	46.0	46.2	46.4	46.7	47.0	47.3	47.9	48.1	48.2	48.7	49.8	50.3	41.7	43.4	45.2	47.0	48.7	50.5	52.3
24	179	44.4	44.6	45.6	46.8	47.1	47.3	47.9	48.0	48.4	48.8	49.1	49.2	50.3	51.5	52.2	41.8	43.9	45.9	48.0	50.0	52.1	54.1
30	206	44.7	46.0	46.5	47.4	47.9	48.3	48.9	49.1	49.9	50.2	50.5	50.8	51.7	52.4	53.5	42.6	45.6	46.9	47.1	51.3	53.5	55.6
36	266	45.5	46.4	47.3	48.4	48.8	49.2	50.0	50.4	50.9	51.1	51.6	51.9	52.9	54.0	55.2	43.2	45.6	48.3	50.4	52.8	55.2	57.6
42	370	46.4	47.0	47.9	49.0	49.4	49.9	50.4	51.2	51.9	52.2	52.6	52.9	53.9	54.9	55.7	44.0	46.4	48.8	51.2	53.7	56.1	58.5
48	432	46.5	47.2	48.0	49.3	49.7	50.1	51.5	52.5	53.0	53.5	54.0	55.3	56.6	57.7	43.0	45.8	48.8	48.6	51.5	54.3	57.2	60.0
54	459	47.2	48.0	48.9	50.0	50.4	50.9	51.4	52.0	53.0	53.5	54.0	54.1	55.9	56.9	57.9	43.9	46.6	49.3	52.0	54.7	57.5	60.2
60	381	47.9	48.3	49.4	50.5	50.9	51.3	50.2	52.9	53.8	54.4	54.7	55.0	56.0	57.5	57.9	44.7	47.4	50.2	52.9	53.7	58.4	61.2
66	245	48.8	49.4	50.0	51.2	52.8	52.4	52.9	53.5	54.6	55.4	55.8	56.1	57.1	58.3	58.9	45.1	47.9	50.7	53.4	56.3	59.1	61.9
72	98	50.2	50.5	50.9	52.7	53.1	53.5	54.7	55.7	56.4	57.0	57.4	57.8	58.6	59.5	61.1	47.0	49.9	52.8	55.7	58.6	61.5	64.4

N: Number of children   Indian Pediatr 1994; 31:377

Table 20: *Showing skin fold percentiles (mm) in relation to breast development and age (years) in girls*

	SMR = 2				SMR = 3				SMR = 4				SMR = 5			
	Tr	Bi	Sb	Sl	Tr	Bi	Sb	Sl	Tr	Bi	Sb	Sl	Tr	Bi	Sb	Sl
Age = 8																
Mean	16.3	8.8	13.8	15.1												
SD	5.4	3.5	6.6	7.3												
N	45	45	45	42												
5	7.2	4.2	5.2	5.0												
15	10.2	5.0	6.0	7.0												
50	17.0	8.0	13.0	14.5												
85	21.0	12.0	21.4	21.0												
90	23.0	14.2	23.6	24.7												
95	23.0	15.8	25.8	27.0												
Age = 9																
Mean	13.8	7.5	11.0	12.3												
SD	6.3	3.8	6.5	7.9												
N	188	188	186	168												
5	6.0	3.0	4.0	4.0												
15	8.0	4.0	5.0	5.0												
50	12.0	7.0	8.5	10.0												
85	21.0	12.0	18.0	21.0												
90	22.0	13.0	20.5	25.0												
95	25.7	15.0	25.0	28.0												
Age = 10																
Mean	13.3	8.2	10.9	12.8	15.1	9.6	13.3	16.3	17.8	11.5	17.7	21.0				
SD	5.5	3.9	6.0	8.2	5.2	4.2	6.9	8.9	5.9	4.5	7.2	8.7				
N	381	378	366	284	129	130	127	111	31	31	30	26				
5	6.0	3.0	5.0	4.0	8.0	5.0	6.0	5.5	9.0	5.5	10.0	8.0				
15	8.0	4.0	6.0	5.0	9.0	5.0	7.0	7.0	13.0	7.5	10.4	13.5				
50	12.0	8.0	8.5	10.0	15.0	9.0	11.0	15.0	17.0	10.0	17.0	19.5				
85	20.0	12.0	18.0	22.0	20.8	14.0	21.0	25.5	24.5	15.5	24.7	29.3				
90	21.0	14.0	20.0	26.0	22.0	15.0	22.0	28.0	25.0	17.0	26.0	31.0				
95	24.0	16.0	22.0	29.9	24.0	17.6	26.7	31.0	27.0	18.0	27.1	32.0				
Age = 11																
Mean	12.5	7.5	9.7	11.4	14.4	9.1	12.4	14.3	16.4	9.9	14.8	17.7				
SD	4.3	3.3	3.5	3.6	5.4	4.5	6.1	7.8	5.9	3.9	6.1	7.8				
N	249	249	249	214	316	316	299	269	115	115	111	102				
5	6.0	3.0	5.0	4.0	7.0	4.0	6.0	5.0	7.0	5.0	7.0	7.1				
15	8.0	4.0	5.0	5.0	8.3	5.0	7.0	7.0	11.0	6.0	8.0	10.0				
50	11.0	6.0	7.0	8.0	14.0	8.0	11.0	12.0	15.0	9.0	13.0	17.0				
85	16.0	10.0	12.0	13.0	20.0	14.0	18.0	23.0	22.0	14.0	22.0	26.0				
90	18.0	11.0	13.0	14.0	22.0	16.0	20.0	25.0	25.0	15.6	24.0	28.0				
95	20.0	14.0	15.0	15.0	24.0	18.0	25.0	28.0	28.0	17.0	25.0	30.0				

Contd.

Table 20: *Showing skin fold percentiles (mm) in relation to breast development and age (years) in girls (Contd.)*

	SMR = 2				SMR = 3				SMR = 4				SMR = 5			
	Tr	Bi	Sb	SI	Tr	Bi	Sb	SI	Tr	Bi	Sb	SI	Tr	Bi	Sb	SI
Age = 12																
Mean	11.6	6.7	8.3	10.4	13.6	8.1	11.4	13.0	16.2	9.6	14.5	16.2	21.2	13.1	21.0	25.0
SD	5.3	3.3	3.7	5.7	5.7	4.1	5.8	7.4	5.3	3.8	5.9	6.7	5.9	4.4	7.1	7.2
N	196	195	151	121	329	329	266	235	244	243	222	204	88	88	86	85
5	6.0	3.0	5.0	5.0	6.0	3.0	6.0	5.0	8.0	5.0	8.0	7.0	10.7	6.4	9.3	13.2
15	7.0	4.0	5.0	6.0	8.0	5.0	7.0	6.1	11.0	6.0	9.0	9.0	16.0	8.0	14.0	16.0
50	11.0	6.0	7.0	9.0	13.0	7.0	10.0	11.0	16.0	9.0	13.0	16.0	21.0	14.0	20.0	25.0
85	16.0	9.0	11.0	15.0	20.0	12.0	16.3	20.0	22.0	14.0	19.9	23.0	27.0	17.0	29.0	33.0
90	19.0	10.6	13.0	18.0	22.0	14.0	18.5	25.0	23.7	15.0	22.0	25.0	28.0	18.0	31.0	33.6
95	22.3	13.3	17.5	20.0	24.6	17.0	25.0	30.0	25.9	16.0	25.0	29.0	29.7	19.7	33.0	35.8
Age = 13																
Mean	9.6	5.8	8.4	8.6	12.5	6.7	10.1	11.7	15.3	8.6	13.8	15.3	20.7	11.8	19.2	21.9
SD	4.0	2.5	3.3	4.6	4.9	3.0	4.9	6.3	5.0	3.4	5.9	7.4	6.1	4.6	7.4	8.9
N	52	52	36	17	241	240	168	140	304	304	267	231	177	177	172	163
5	4.6	2.6	5.0	3.8	6.0	3.0	6.0	5.0	8.0	4.0	7.0	6.0	12.0	6.0	8.0	9.0
15	6.0	3.7	5.0	4.4	8.0	4.0	7.0	6.0	10.0	6.0	9.0	8.0	14.4	7.0	12.0	12.0
50	9.0	5.0	7.0	7.0	12.0	6.0	9.0	10.0	15.0	8.0	12.0	14.0	20.0	11.0	19.0	22.0
85	14.0	8.4	11.0	12.6	17.0	9.0	13.0	16.2	20.0	12.0	18.0	22.0	27.0	16.0	28.0	31.0
90	14.0	9.0	12.0	14.2	19.0	10.0	15.0	19.1	22.0	13.0	21.0	25.0	28.0	18.0	28.0	33.0
95	16.9	10.0	14.8	16.6	22.0	12.0	19.3	24.1	24.0	15.0	25.0	30.0	30.2	21.0	30.5	38.0
Age = 14																
Mean					12.2	6.5	10.6	10.7	14.9	8.1	12.7	13.6	19.9	11.5	17.8	19.4
SD					4.5	2.5	4.0	3.8	5.1	3.4	4.7	5.6	5.7	4.9	6.8	8.6
N					143	143	71	42	263	263	216	170	189	189	181	166
5					6.0	3.0	6.0	6.0	8.0	4.0	6.8	7.0	12.0	6.0	9.0	8.0
15					8.0	4.0	7.0	7.0	10.0	5.0	8.0	8.0	14.0	7.0	10.0	10.0
50					12.0	6.0	10.0	10.0	14.0	8.0	12.0	12.0	20.0	11.0	18.0	18.0
85					17.0	9.0	14.0	14.9	20.0	12.0	17.0	20.0	25.0	16.0	25.0	30.0
90					18.8	10.0	16.0	16.0	22.0	13.0	18.8	21.0	27.0	18.0	27.0	32.0
95					20.0	12.0	19.5	18.0	24.0	14.0	21.9	25.7	29.6	20.0	31.0	35.0
Age = 15																
Mean					11.5	6.9	11.1	10.9	15.2	8.9	14.4	15.2	19.1	10.3	17.5	18.9
SD					3.8	2.7	3.3	3.8	5.5	3.5	6.1	7.4	6.1	4.5	7.3	8.9
N					71	71	56	35	141	141	119	89	187	187	179	178
5					6.0	4.0	7.0	5.7	9.0	5.0	8.0	6.4	11.0	4.0	8.0	8.0
15					7.5	4.5	8.0	7.0	10.0	6.0	9.0	9.0	13.0	6.0	10.0	10.0
50					11.0	6.0	10.0	10.0	15.0	8.0	13.0	13.0	18.0	9.0	16.0	17.5
85					16.0	9.0	15.0	16.0	21.0	12.0	19.3	22.8	27.0	15.0	25.0	28.0
90					16.0	10.0	15.0	16.6	22.0	13.0	22.0	25.2	28.0	16.0	26.2	31.3
95					18.0	12.5	18.0	18.0	24.1	15.1	26.0	29.0	30.0	18.0	31.0	37.2

Contd.

Table 20: *Showing skin fold percentiles (mm) in relation to breast development and age (years) in girls (Contd.)*

	SMR = 2				SMR = 3				SMR = 4				SMR = 5			
	Tr	Bi	Sb	SI	Tr	Bi	Sb	SI	Tr	Bi	Sb	SI	Tr	Bi	Sb	SI
Age = 16																
Mean					11.5	6.3	10.8	11.4	14.5	7.8	15.3	14.8	19.4	10.2	16.4	16.3
SD					3.9	2.9	3.7	8.6	5.2	3.8	5.8	7.4	6.5	5.0	8.0	7.6
N					34	34	24	11	89	89	55	38	146	146	125	120
5					6.7	3.0	7.0	6.0	8.0	3.0	7.7	7.0	10.0	5.0	7.0	6.0
15					7.0	4.0	7.5	6.5	9.0	4.0	9.0	8.0	12.0	5.8	9.0	8.0
50					11.0	5.0	10.0	8.0	14.0	7.0	15.0	14.0	19.0	9.0	14.0	15.0
85					15.1	10.0	13.6	15.0	20.0	11.8	21.0	21.5	26.0	15.0	25.0	24.0
90					16.0	10.0	16.1	18.0	22.0	13.0	23.2	24.3	27.0	17.0	27.0	25.1
95					17.7	11.4	17.9	26.5	24.0	14.0	28.0	29.3	30.8	20.8	31.8	29.1
Age = 17																
Mean									14.9	7.7	14.9	16.3	19.4	10.1	15.8	17.0
SD									6.2	4.4	6.1	6.4	5.1	5.0	6.6	8.3
N									44	44	28	25	60	60	46	46
5									8.0	4.0	8.0	8.2	12.0	5.0	7.3	7.0
15									10.0	4.5	9.0	9.6	14.0	5.9	10.0	9.0
50									13.0	6.0	14.0	16.0	19.0	9.0	14.0	15.0
85									21.0	10.0	19.9	21.0	25.2	15.2	24.0	25.0
90									23.0	11.7	21.6	22.2	26.1	18.0	26.0	28.0
95									30.0	20.0	31.0	35.0	29.0	20.0	28.0	32.8

Tr: Tricep, Bi: Bicep, Sb: subscapular and Si: Suprailiac
Indian Pediatr 2001;38:1217–35

Table 21: *Showing skin fold (mm) percentiles in relation to genital development and age (years) in boys*

	SMR = 2				SMR = 3				SMR = 4				SMR = 5			
	Tr	Bi	Sb	SI	Tr	Bi	Sb	SI	Tr	Bi	Sb	SI	Tr	Bi	Sb	SI
Age = 8																
Mean	10.1	7.1	8.0	8.6	12.5	12.5	12.5	12.5	12.5	12.5	12.5	12.5	12.5	12.5	12.5	12.5
SD	3.8	4.1	5.3	4.7												
N	27	27	27	9												
5	6.0	3.0	4.3	5.0												
15	6.9	3.9	5.0	5.0												
50	9.0	6.0	6.0	7.0												
85	13.0	9.2	11.3	12.2												
90	13.8	12.6	14.4	14.2												
95	17.1	15.7	19.2	16.6												

Contd.

Table 21: *Showing skin fold (mm) percentiles in relation to genital development and age (years) in boys (Contd.)*

	SMR = 2				SMR = 3				SMR = 4				SMR = 5			
	Tr	Bi	Sb	SI	Tr	Bi	Sb	SI	Tr	Bi	Sb	SI	Tr	Bi	Sb	SI
Age = 9																
Mean	8.9	5.8	6.6	7.9												
SD	3.9	2.6	3.1	4.8												
N	102	102	102	43												
5	5.0	3.0	4.0	4.0												
5	6.0	3.0	5.0	4.0												
50	8.0	5.0	6.0	6.0												
85	12.0	8.0	8.0	11.7												
90	12.9	9.9	10.0	12.8												
95	15.0	11.9	12.0	16.9												
Age = 10																
Mean	10.7	6.4	8.0	9.5	9.1	5.8	7.5	10.8								
SD	5.3	3.9	4.7	6.7	4.3	3.4	4.6	7.5								
N	254	254	254	153	73	73	71	33								
5	5.0	3.0	4.0	4.0	4.6	2.0	4.0	5.0								
15	6.0	3.0	5.0	5.0	5.0	3.0	4.0	5.0								
50	9.0	5.0	6.0	7.0	8.0	5.0	6.0	7.0								
85	16.0	10.0	12.0	15.0	14.2	9.0	11.5	18.8								
90	18.7	12.7	15.0	19.0	15.0	10.0	13.0	24.4								
95	22.0	15.0	19.0	21.4	16.0	12.8	16.5	28.0								
Age = 11																
Mean	10.9	6.3	8.7	11.2	10.2	6.0	8.8	11.3	9.8	5.4	9.4	11.1				
SD	5.0	3.5	5.5	8.0	5.1	3.9	5.5	8.4	5.3	3.0	5.2	7.3				
N	467	467	467	302	152	152	152	115	28	28	28	27				
5	5.0	3.0	4.0	4.0	5.0	3.0	5.0	5.0	5.0	3.0	5.0	6.0				
15	6.0	3.0	5.0	5.0	6.0	3.0	5.0	6.0	6.0	3.0	6.0	6.0				
50	10.0	5.0	7.0	8.0	8.5	5.0	7.0	8.0	8.0	4.5	7.5	8.0				
85	16.0	10.0	14.0	20.0	15.4	8.3	12.0	16.9	14.0	8.0	14.8	15.0				
90	18.0	12.0	16.0	24.9	16.0	10.9	15.0	25.0	15.0	9.0	18.6	20.2				
95	21.0	13.0	20.7	28.0	20.5	12.9	20.4	28.9	21.5	11.0	20.0	28.7				
Age = 12																
Mean	11.4	6.4	9.1	12.0	10.1	5.7	8.8	11.5	9.3	5.1	8.5	10.9				
SD	5.3	3.6	6.1	8.3	4.7	3.3	5.2	7.8	4.1	2.6	3.9	6.3				
N	488	488	482	382	350	350	350	269	156	156	156	149				
5	5.0	3.0	4.0	4.0	5.0	2.0	5.0	5.0	5.0	2.0	5.0	5.0				
15	7.0	3.0	5.0	5.0	6.0	3.0	5.0	5.0	6.0	3.0	6.0	6.0				
50	10.0	5.0	7.0	9.0	9.0	5.0	7.0	9.0	8.0	4.0	7.0	8.0				
85	18.0	10.0	15.0	20.0	15.0	8.0	13.0	20.0	14.0	7.0	12.0	18.0				
90	20.0	11.3	17.0	25.9	16.0	10.0	16.0	25.0	15.0	8.5	13.0	20.2				
95	22.0	14.0	22.0	29.0	20.0	12.0	21.6	28.6	17.0	10.0	17.3	24.0				

Contd.

Table 21: *Showing skin fold (mm) percentiles in relation to genital development and age (years) in boys (Contd.)*

	SMR = 2				SMR = 3				SMR = 4				SMR = 5			
	Tr	Bi	Sb	SI	Tr	Bi	Sb	SI	Tr	Bi	Sb	SI	Tr	Bi	Sb	SI
Age = 13																
Mean	11.8	6.6	9.7	11.9	10.0	5.6	8.2	11.0	10.1	5.5	9.5	11.6	10.5	6.0	10.4	13.2
SD	5.8	4.0	7.4	9.3	4.7	3.0	4.7	7.0	4.8	3.0	5.5	7.8	5.2	3.3	6.3	8.4
N	267	267	267	220	422	422	421	318	415	415	415	366	79	79	79	78
5	6.0	3.0	4.0	4.0	5.0	2.0	4.0	5.0	5.0	3.0	5.0	5.0	5.0	3.0	5.0	6.0
15	7.0	4.0	5.0	5.0	6.0	3.0	5.0	5.0	6.0	3.0	6.0	6.0	6.0	3.7	6.0	7.0
50	10.0	5.0	7.0	9.0	8.0	5.0	7.0	8.0	9.0	5.0	8.0	9.0	9.0	5.0	8.0	10.0
85	17.1	10.0	14.1	20.0	15.0	8.0	12.0	17.0	14.0	8.0	12.0	18.0	15.0	8.3	14.0	20.0
90	20.0	12.0	20.0	27.1	16.0	10.0	14.0	21.0	16.0	9.0	15.0	23.5	18.0	10.0	16.2	23.6
95	24.0	15.0	29.4	33.1	20.0	12.0	18.0	27.0	20.3	11.3	21.0	30.0	20.0	12.0	21.1	31.0
Age = 14																
Mean	11.5	6.4	10.0	13.1	10.0	5.5	8.7	10.7	9.5	5.2	9.5	11.3	11.2	6.2	11.2	14.0
SD	5.6	3.3	6.0	7.3	5.5	3.6	5.4	6.7	4.7	2.9	5.4	7.1	5.7	3.7	6.3	8.6
N	79	79	78	65	271	271	268	198	524	524	524	439	314	314	314	307
5	5.0	3.0	4.0	5.0	4.0	2.0	4.4	4.0	5.0	2.0	5.0	5.0	5.0	3.0	6.0	5.0
15	6.0	3.7	5.0	6.0	5.0	3.0	5.0	5.0	6.0	3.0	6.0	6.0	6.0	3.0	7.0	7.0
50	10.0	5.0	8.0	11.0	8.0	4.0	7.0	8.0	8.0	4.0	8.0	9.0	9.0	5.0	9.0	11.0
85	17.3	10.0	15.0	20.4	15.0	8.5	12.0	18.0	13.0	7.0	12.0	16.0	16.0	9.0	17.0	22.0
90	20.0	11.0	16.6	23.2	17.0	10.0	15.0	21.3	15.7	9.0	15.0	20.0	18.0	11.0	19.7	26.4
95	21.1	13.1	23.2	25.8	22.5	13.5	20.0	26.2	20.0	11.0	21.0	26.0	22.4	14.0	25.0	32.0
Age = 15																
Mean					10.5	5.5	10.0	11.6	10.5	5.5	10.8	12.6	11.0	6.0	11.6	14.7
SD					6.1	4.0	6.6	7.7	5.3	3.2	6.5	7.7	5.2	3.3	6.6	9.1
N					100	100	100	65	391	391	391	268	457	457	457	433
5					5.0	3.0	5.0	5.0	5.0	3.0	5.0	5.0	5.0	3.0	6.0	6.0
15					6.0	3.0	6.0	6.0	6.0	3.0	6.0	6.1	6.0	3.0	7.0	7.0
50					9.0	4.0	8.0	9.0	9.0	4.0	8.0	10.0	10.0	5.0	9.0	12.0
85					14.2	8.0	13.2	18.2	15.0	8.0	16.0	20.0	16.0	9.0	18.0	24.2
90					19.1	10.0	17.1	21.6	18.0	10.0	20.0	24.3	18.0	10.0	21.0	28.0
95					25.1	13.1	24.1	28.2	20.5	13.0	24.5	29.3	22.0	13.0	25.0	32.4
Age = 16																
Mean					10.3	5.4	10.2	12.9	9.8	5.0	10.9	11.6	10.3	5.6	11.6	14.2
SD					4.8	3.3	6.2	9.3	5.3	3.3	6.9	6.6	4.6	3.1	5.6	8.0
N					43	43	43	35	182	182	182	103	376	376	376	331
5					5.0	2.0	5.0	4.7	5.0	2.1	6.0	6.0	5.0	3.0	6.0	6.0
15					6.0	3.0	6.0	6.0	5.2	3.0	6.0	6.3	6.0	3.0	7.0	7.0
50					9.0	4.0	8.0	10.0	8.0	4.0	9.0	10.0	9.0	5.0	10.0	12.0
85					15.0	9.4	16.8	20.9	15.0	7.0	13.9	16.0	15.0	8.0	16.0	23.0
90					17.6	11.0	18.0	23.8	16.9	8.0	19.8	20.0	16.0	9.0	18.0	26.0
95					19.8	12.0	23.5	35.9	20.0	11.0	26.9	27.8	20.0	11.0	22.0	30.0

Contd.

Table 21: *Showing skin fold (mm) percentiles in relation to genital development and age (years) in boys (Contd.)*

	SMR = 2				SMR = 3				SMR = 4				SMR = 5			
	Tr	Bi	Sb	SI	Tr	Bi	Sb	SI	Tr	Bi	Sb	SI	Tr	Bi	Sb	SI
Age = 17																
Mean									9.8	5.0	10.7	11.8	10.4	5.5	12.4	14.8
SD									4.8	2.7	5.5	7.4	4.7	3.0	5.8	8.8
N									58	58	58	37	241	241	241	192
5									5.0	2.9	6.0	6.0	5.0	3.0	6.0	6.0
15									6.0	3.0	7.0	7.0	6.0	3.0	7.0	7.0
50									8.5	4.0	9.0	10.0	9.0	5.0	10.0	12.0
85									15.0	7.0	14.0	15.6	15.0	8.0	18.0	25.0
90									16.0	8.3	17.0	19.4	16.0	9.0	20.0	27.9
95									17.5	10.3	22.6	25.0	18.0	10.0	25.0	32.0

Tr: Tricep, Bi: Bicep, Sb: subscapular and Si: Suprailiac

Indian Pediatr 2001;38:1217–35

Table 22: *Percentage distribution of boys according to genital development at each age point*

Age (yr)		Genital stages							
		2		3		4		5	
		N	%	N	%	N	%	N	%
8.0	233	2	0.86						
8.5	213	10	4.69	3	1.41				
9.0	285	34	11.93	13	4.56				
9.5	286	51	17.83	6	2.10	1	0.35		
10.0	386	80	20.73	24	6.22	3	0.78		
10.5	556	127	22.84	33	5.94	3	0.54		
11.0	580	185	31.90	54	9.31	3	0.52		
11.5	662	241	36.40	79	11.93	22	3.32		
12.0	677	265	39.14	115	16.99	24	3.55	3	0.44
12.5	770	275	35.71	197	25.58	76	9.87	3	0.39
13.0	695	215	30.94	214	30.79	139	20.00	10	1.44
13.5	736	142	19.29	241	32.74	227	30.84	42	5.71
14.0	680	76	11.18	225	33.09	272	40.00	79	11.62
14.5	677	48	7.09	157	23.19	289	42.69	166	24.52
15.0	584	19	3.25	88	15.07	249	42.64	220	37.67
15.5	482	7	1.45	57	11.83	198	41.08	217	45.02
16.0	409	1	0.24	32	7.82	158	38.63	218	53.30
16.5	344	1	0.29	27	7.85	92	26.74	224	65.12
17.0	266			7	2.74	72	27.07	187	70.30
17.5	165			6	3.64	24	14.55	135	81.82
18.0				2	2.35	16	18.82	67	78.82
Total		1779		1580		1868		1571	
Age mean (yr)		11.3		12.8		14.1		16.4	
SD		1.4		1.5		1.3		1.3	
SE		0.03		0.04		0.03		0.03	

N: Number of subjects

SD: Standard deviations, SE: Standard error.

Table 23: *Percentage distribution of axillary hair, pubic hair and facial hair presence in relation to different genital development stages*

Genital development stages	Axillary hair		Pubic hair		Facial hair	
	N	%	N	%	N	%
2	109	6.13	1059	59.53	168	9.44
N = 1779						
3	855	54.11	1532	96.96	797	50.44
N = 1580						
4	1833	98.13	1861	99.63	1763	94.38
N = 1868						
5	1569	99.87	1569	99.87	1566	99.68
N = 1571						
Mean age (yr)	14.9		14.2		14.8	
SE	0.02		0.24		0.03	

N: Number of subjects; SE: Standard error

Table 24: *Percentage distribution of girls according to breast development at each age point*

Age (yr)	N	Breast stages							
		2		3		4		5	
		N	%	N	%	N	%	N	%
8.0	240	13	5.42						
8.5	288	25	8.68						
9.0	348	45	12.93	6	1.72				
9.5	398	110	27.64	8	2.01	1	0.25		
10.0	451	166	36.81	40	8.87	1	0.22		
10.5	437	182	41.65	56	12.81	18	4.12	1	0.22
11.0	496	226	45.56	123	24.80	30	6.05	2	0.40
11.5	483	184	38.10	174	36.02	62	12.84	4	0.83
12.0	434	142	32.72	159	36.64	83	19.12	17	3.92
12.5	489	92	18.81	206	42.13	133	27.20	44	9.00
13.0	455	56	12.31	172	37.80	164	36.04	59	12.97
13.5	454	31	6.83	157	34.58	163	35.90	101	22.25
14.0	388	14	3.60	120	30.85	164	42.16	90	23.14
14.5	350	2	0.57	84	24.00	166	47.43	98	28.00
15.0	291	3	1.03	60	20.62	121	41.58	107	36.77
15.5	203			43	21.18	73	35.96	87	42.86
16.0	176			28	15.91	64	36.36	84	47.73
16.5	182			26	14.29	64	35.16	92	50.55
17.0	116			5	4.31	20	17.24	91	78.44
Total		1291		1482		1355		834	
Age mean (yr)		10.2		19.6		13.5		15.5	
SD		1.3		1.6		1.6		1.5	
SE		0.04		0.04		0.04		0.05	

N: Number of children
SD: Standard deviations, SE: Standard error.

Table 25: *Percentage distribution of axillary hair, pubic hair and menarche appearance in relation of different breast development stages*

Breast development stages	Axillary hair		Pubic hair		Menarche attained	
	N	%	N	%	N	%
2 N = 1291	309	23.93	284	22.00	57	4.42
3 N = 1482	1365	92.11	1370	92.44	752	50.74
4 N = 1355	1340	98.89	1338	98.75	1223	90.26
5 N = 834	834	100.00	831	99.64	805	96.52
Mean age (yr)	13.6		13.6		12.6	
SE	0.03		0.03		0.04	

N: Number of subjects; SE: Standard error

Table 26: *Showing comparative figures for initiation of sexual development in the US and Indian children*

Wu et al 2002; USA	N=1168	Age = 8–16 yr	SMR= B2 Afro-American 9.5 yr Whites 10.3 yr	Menarche 12.2 yr 12.6 yr	Pubic hair 2 9.5 yr 10.6 yr	Girls
Agarwal et al 1992 INDIA	N= 1291	8–17 yr	10.2 yr	12.6 yr	22% had hair at B₂	Girls
Herman- Giddens et al 2001 USA	2114	8–19 yr	Afro-American 9.5yr Whites 10.1 yr	– –	11.2 yr 12.0 yr	Boys
Agarwal et al 1992 INDIA	1779	8–18 yr	11.3 yr	–	59.3%	Boys

- Herman-Giddens MA, Wang L, Koch G. Secondary sexual characteristics in boys: estimates from the National Health and Nutrition Examination Survey III, 1988–1994. *Arch PediatrAdolesc Med* 2001;155(9):1022–28.

- WuT. Mendola P and Buck G. Ethnic differences in the presence of secondary sex characteristics and menarche among US girls: The Third National Health and Nutrition Examination Survey, 1988–1994. Pediatrics 2002;110: 752–7.

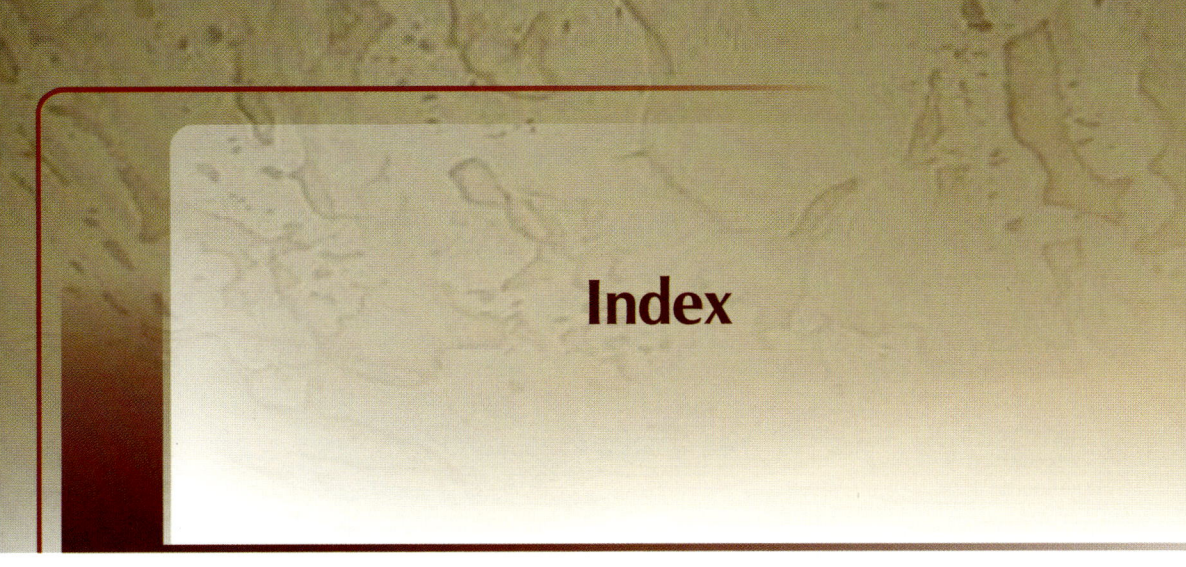

Index